Allied Medicine in the Great War

Jennifer S. Lawrence

Allied Medicine in the Great War

The Medical Front and the People Who Fought

BLOOMSBURY ACADEMIC
LONDON • NEW YORK • OXFORD • NEW DELHI • SYDNEY

BLOOMSBURY ACADEMIC
Bloomsbury Publishing Plc
50 Bedford Square, London, WC1B 3DP, UK
1385 Broadway, New York, NY 10018, USA
29 Earlsfort Terrace, Dublin 2, Ireland

BLOOMSBURY, BLOOMSBURY ACADEMIC and the Diana logo
are trademarks of Bloomsbury Publishing Plc

First published by Red Globe Press 2019
Reprinted by Bloomsbury Academic 2023

Copyright © Jennifer S. Lawrence, 2019

Jennifer S. Lawerence has asserted her right under the Copyright,
Designs and Patents Act, 1988, to be identified as the author of this work.

For legal purposes the Acknowledgements on p.159 constitute
an extension of this copyright page.

All rights reserved. No part of this publication may be reproduced or
transmitted in any form or by any means, electronic or mechanical,
including photocopying, recording, or any information storage or retrieval
system, without prior permission in writing from the publishers.

Bloomsbury Publishing Plc does not have any control over, or responsibility for,
any third-party websites referred to or in this book. All internet addresses given
in this book were correct at the time of going to press. The author and publisher
regret any inconvenience caused if addresses have changed or sites have
ceased to exist, but can accept no responsibility for any such changes.

A catalogue record for this book is available from the British Library.

A catalog record for this book is available from the Library of Congress.

ISBN: HB: 978-1-3520-0446-5
PB: 978-1-3520-0419-9
ePDF: 978-1-3520-0420-5
ePUB: 978-1-3503-0742-1

To find out more about our authors and books visit
www.bloomsbury.com and sign up for our newsletters.

To my father, who taught me to love history

Contents

List of Illustrations   xi

Introduction   xiii

1 The Great War   **1**
The Start of War   3
US Entrance into the War   7
Logistics of War   11
Medical Advancements Prior to World War I   13
Technology of the War   14
Soldiers in the War   19

2 Allied Medical Innovations   **25**
Wounds   29
Orthopedics   34
Transfusions   38
X-Rays   40
Plastic Surgery   44
Conclusion   50

3 Medical Personnel — 53
Medical Corps—General Experiences — 54
Volunteering — 60
Post-US Entrance — 65
Stretcher-Bearers — 68
Nursing — 70
Staying the Course — 73
Influenza — 74
Home-Front Treatment — 77
Race — 79
Bureaucracy — 80
Conclusion — 81

4 Soldiers and the Medical Front — 83
Reacting to the Wounded — 84
Prevention of Problems — 86
Surgery — 89
Gas — 91
Camaraderie — 95
Soldiers and Noncombat Injuries — 98
Aftercare — 101
Psychological Care — 104
Conclusion — 106

5 Effects of the Medical Front in the Great War — 109
Veterans and Medical Care after the War — 112
Orthopedics after the War — 116
Plastic Surgery after the War — 117
War Medicine in Civilian Practice — 122
Influenza — 126
Medical Bureaucracy and Public Health — 126
Shortcomings in Medicine — 131
Conclusion — 134

Conclusion — 139

Notes 145

Acknowledgments 159

Bibliography 163
Primary Sources 163
Secondary Sources 167
Journals 169

Index 171

List of Illustrations

Illustration 2.1	Demonstration of orthopedic splints and the use of pulleys in treating a fracture patient	36
Illustration 3.1	Nurses and a doctor tend to a wounded soldier	64
Illustration 3.2	British Army medical collection point	69
Illustration 4.1	Various styles of gas masks	92
Illustration 4.2	Autograph book from a hospital	103
Illustration 5.1	A French soldier before having plastic surgery performed	118
Illustration 5.2	A French soldier after having plastic surgery performed	119

Introduction

On June 28, 1914, a choreographed visit by an imperial official went horribly awry. Archduke Franz Ferdinand and his wife toured the city of Sarejevo in Bosnia-Herzegovina. After an aborted and unsuccessful attempt on his life in the morning, a terrorist group would find its fortunes changed by the afternoon. Both Ferdinand and his wife, Sophie, were shot and killed by Gavrilo Princip, a 19-year-old who hoped to strike a blow for the cause of a greater Serbia. As the archduke was the heir to the throne of the once-mighty Austro-Hungarian Empire, his uncle, the sitting emperor, wanted answers. Many of the diplomats the emperor was surrounded with wanted war. They sought to crush their hated enemy of Serbia and saw this as the consummate opportunity. Once they pursued that policy to the brink, the dominos of the alliances and diplomatic relations of the European countries began to fall. The war would envelop numerous countries and empires alongside countless families and individuals over the next four years.

As the war ground on, hundreds of thousands and then millions of soldiers from across Europe and the world became enveloped in the conflict. The scale of these military forces was unprecedented. Armies had to develop ways to provide weapons, food, and supplies to their ever-growing numbers. They also had to support continued efforts to bring aid and medical attention to those caught up in its vortex.

War and medicine have intertwined themselves deeply throughout human conflict. In his article "The Medicalization of War—The Militarization of Medicine," Mark Harrison noted that even the metaphors of each profession are filled with references to one another. "Germs are the 'unseen enemy'... the 'war against disease'... the much vaunted 'surgical strike.'"[1] The sheer numbers of wounded soldiers who needed treatment ensured that World War I would provide medicine a unique opportunity to advance its knowledge exponentially in only a few years. The repetition of various treatments allowed doctors to better judge which treatments offered the wounded the most benefits and which treatments should be discarded. The war also provided a time for the medical field to put into practice several of the theories it had developed since the turn of the century. It would also give its practitioners opportunities to work together to help these patients in a fashion that often did not exist in the civilian medical world.

The governments at war were beholden to "construct medical organizations far larger and more centralized than anything conceivable in peacetime."[2] This was not simply one country's first foray into providing medical care on this scale; it was every country's first foray into medical care on this scale. Not surprisingly, medicine became increasingly bureaucratic during the war. Part of this was due to the logistics necessary to implement medical care to this many patients. But organization also dominated the field itself, and specialties such as maxillofacial and orthopedics became increasingly important. Patients with abdominal, head, and chest wounds were assigned special teams as well.[3] This emphasis on specialization transplanted itself into civilian medicine after the war as an increasing number of physicians took up a specialty.

What made all of this possible, though, were the people who provided this care—doctors, nurses, stretcher-bearers, ambulance drivers. Without them the best science could never be delivered to the patient. The patients too found themselves caught up in the medical front of the war as so many faced their own turn at being treated. The medicine that emerged from these four years went on to reshape civilian medical practice in the decades after the war.

This book is designed to introduce the reader to the overall experience of medical practice on the Allied Western Front during World War I, with particular focus on the operations of Great Britain and the United States. It seeks to provide in a single volume an overview of this topic, from the medicine itself, with the numerous innovations that were made during the war, to the experience of doctors, nurses, and soldiers who participated in this "front" of the war. It integrates this issue of medicine with the larger conduct of the war. It seeks to do this in a readable and accessible form that also highlights how this war's medical elements had long-term consequences on the field of medicine and for future patients.

The book deliberately does not address the actions on other fronts of the war on which other Allied armies fought. The largest other front was the Eastern Front, which engaged mostly Russian troops. The organization of the Russian military and subsequently its medical corps suffered throughout the war from shortages and comprehensive training. Although many medical personnel on the Western Front spoke English or had a knowledge of French and could as such communicate across national lines, the knowledge of the Russian language and the smaller numbers of Russian medical publications meant that Russian medical approaches did not have as much influence on the Western theater of the war and its medical practices.

The first chapter of this book outlines how the war came to be and how it was fought, for those who may be unfamiliar with these developments. It describes the conditions under which soldiers were expected to fight and how these conditions affected the number of casualties. It also explains how the medical response to the war unfolded. Soldiers found themselves immediately caught up in fighting that they did not always understand and had little control over. Numerous new military technologies were brought to bear on the battlefields. The second chapter explores the medical science of the war. It addresses the treatments that were given for various wounds and how these treatments were a furthering of medical know-how during the war. A number of innovations and new practices meant the wounded received the best medical care at the time, care that was truly cutting-edge for the era. World War I was a time for specialists to come into their own, offering new treatments that would become standardized and provide better recovery for the patients.

The third chapter examines the medical personnel themselves—the doctors, nurses, and stretcher-bearers. It explains what life was like for these individuals during the war, how they coped with the pressure of their duties, and how they experienced the war. It also illuminates how they found solace and peace and even times of joy in these years. Many of the medical personnel in the war came into it via volunteering efforts. The medical side of the war became more bureaucratic as the war progressed, and the medical personnel found themselves on a learning curve throughout the war as they adopted more effective ways to approach their patients. The fourth chapter explores the medical front from the point of view of the patient. Most soldiers had some sort of contact with the medical front during the war, either because of their own wounds or through the wounding of a friend or fellow soldier. Utilizing published diaries and memoirs, Chapter 4 observes how these soldiers viewed the medical care they received and the personnel who delivered it. It also considers how soldiers viewed getting wounded in this war and what they came to expect from their involvement with the medical front. The last chapter tackles what these medical innovations and practices meant outside of the military. How did what doctors and nurses learned in the war translate to civilian society? How did others benefit from the new science and specializations? How would governments and patients fare after the war? What was the impact on public health? How did medicine change because of the war?

Within these divisions, this book also addresses the US Medical Department's participation in the war. Quite often, most existing volumes on medicine in the war focus fairly exclusively on the British and French roles in the medical front, but little exists that details the US role and contribution. Although the United States was not a combatant for the entire conflict, the government and military did build on other Allied experiences in organizing its medical service, and the work of the medical staff from the United States played an important role in spreading the impact of wartime medicine to civilian hospitals when the war was over.

A number of insightful volumes have been produced in the past few years, detailing various specific aspects of medicine in the Great War. *Wounded*, by Emily Mayhew (2013), chooses to focus on select individuals, telling its tale of the effects of battle away from the front lines as patients faced their wounds and treatment. Mayhew tackles these examples with a literary approach, arousing a visceral and sympathetic reaction by the reader. Mark Harrison's (2010) massive volume on the British medical experience is exhaustive in its detail. *Surgery in War 1914–18*, edited by Thomas Scotland and Steven Heys (2013), takes a more scientific lens to medical topics. Essays on various specialties written by medical doctors appear in it, highlighting the developments and innovations in those specializations. For a person interested in a highly detailed volume, this is a must. It does not address, however, how medicine was practiced in other areas of battle such as the Eastern Front or the Middle East or Africa. But its value is in how it highlights this incredibly dynamic theater of the war and makes concrete links to the impact of war in civilian society. It further humanizes this war for readers who are unfamiliar with what happened after a battle ended. Other volumes are appearing that address the issue of the wounded's return to ordinary life, such as the book *War's Waste*, by Beth Linker (2011), which addresses the problems and successes of follow-up care for amputee veterans after the war; or *Broken Men*, by Fiona Reid (2014), which explores how shell-shocked soldiers fared in the next phase of their lives.

In recent years, as the centenary of the Great War occurred, a spate of memoirs from both soldiers and medical personnel have appeared as well. These works provide greater context and understanding of the experience of those dealing with the medical front of the war. Some of these texts had earlier been printed in the 1930s but had long fell to being out of print and difficult to obtain. The resurrection of these books has been a valuable source for World War I historians. In other cases, diaries had lain unpublished until the interest in this topic grew in the early twenty-first century, triggering publication of these new works.

Although there are some accounts from doctors who practiced during the war, there have been numerous works on the nursing profession during the war. These come both as primary sources originating in wartime letters and diaries by the nurses themselves, and then as secondary works by historians who have presented compelling accounts of how nursing evolved during the years and what nurses faced in their various capacities and roles throughout the war. Christine Hallett has humanized the work of the nurses in the war in two of her volumes. In *Containing Trauma: Nursing Work in the First World War* (2009) and *Veiled Warriors: Allied Nurses of the First World War* (2014), Hallett disabuses readers of the simple myth of the romantic nurse tending to a patient that emerged from the war, and instead describes the daily work of the nurses and their all-too-human reactions to both the war and how they were treated by other medical personnel, the military, and the patients.

All of these volumes and others like them are most instructive in highlighting aspects of the medical front of World War I, but their specific focus also means that it is easy to lose sight of the big picture of the medical front in the war. This work is not meant to comprehensively cover everything that transpired on the medical side of this war—such would be an impossible task. Hopefully, though, this book fills the gap between many narrowly focused works on medicine in the war and ones that are so broad as to leave the reader more interested in the Great War, scrounging for mention of its medical aspects.

1

The Great War

The Great War. The war to end all wars. Such brief statements are accurate and simple, but these also speak to the shock and disillusionment that World War I created. People needed these phrases to reflect the momentous nature of the conflict that upended their worlds. The concept of war changed from one focused on patriotism and heroism, though it was always darker than that. Instead, it became viewed as a man-made encounter that fosters brutality, suffering, and inhumanity. It was a beast that seemed to feed on itself, growing more difficult to stop as the months passed on. At no time was this ever more true than during the years of 1914 to 1918, when the Great War dominated world affairs. It was impossible to contemplate another war after this one.

This global conflagration burned up seemingly an entire generation of men. The war that was supposed to be so short, and was entered so hurriedly, dragged on for over four harrowing years. Approximately 65 million soldiers fought in World War I. This was an unheard-of sum at the time, and is difficult to grasp today. To put this in context, consider that the largest Western war in the decades immediately preceding World War I was the US Civil War. In that war, over 2.5 million Americans fought and over 600,000 died. More than 20-fold of that number fought in World War I.

In the Great War, nations faced the reality of industrial warfare writ large. The past European conflicts—the Austro-Prussian War, the

Franco-Prussian War, or even the imperial fight of the British in the Boer War—each gave their own tantalizing hints of what industrial military might could mean. There had been new weapons: repeating rifles, accuracy at longer ranges, larger artillery shells, early versions of machine guns. But these conflicts bore more similarities to the military machinations of the Napoleonic era than to what warfare would mean in the twentieth century with the advanced weaponry that industrialization hath wrought.

In the years preceding the war, the possibility of armed conflict shadowed European nations such as France, Germany, and Russia. The system of alliances and balances of power that had held major warfare in check since the post-Napoleonic period began to teeter in the 1890s. This was in no small part due to the personalities of leaders and the, at times, moribund state of bureaucracy and government in several of the imperial governments. Paranoias, suspicions, and vacillations became the hallmark of government policy in more than one country. Diplomatic gamesmanship among the major powers of Europe continued through the quarter century prior to the start of the Great War, but with no clear hint that a full-out war was predetermined. Instead, outwardly, many of these countries diplomatically sought to avoid war. However, at the same time, Germany was building a naval fleet and new artillery guns that would be used in any coming conflict. Britain perfected its first dreadnought ship, revolutionizing naval vessels for all time and rendering irrelevant any combat ship design from the past. The arms race that occurred in the early 1900s did not happen by accident.

Although governments are often likened to monolithic institutions, the reality is that policy can be influenced by individuals. The personalities of some key people helped influence the eventual start and enlargement of the war. There were also psychological splits to be aware of between policy and personality. Germany was simultaneously portrayed as a warmonger in the years before 1914 and also praised for its love of peace. It was not a foregone conclusion that Germany would support a war in 1914.

The paranoia of Kaiser Wilhelm II, Germany's monarch since 1888, contributed to the German perspective of its status in world affairs. The idea that Germany was encircled by its enemies held sway in the early twentieth century as Germany pointed toward both Russia and France as potential future opponents. The kaiser did not think he was respected as much as he deserved to be by fellow world leaders. Germany sought to increase its colonial holdings, taking a foothold in Samoa and eyeing other ports in the Pacific. Anxiety over German attempts to gain an Atlantic port in North Africa caused international incidents in 1905 and 1911.

In Austro-Hungary, the chief of the General Staff, General Franz Conrad von Hötzendorf, dreamed of a European map without Serbia on it. Austria, he believed, needed to take over that problem territory, and he envisioned a small war should be able to accomplish that.

People spoke of a potential "Great European War" coming in the future, but for civilians that seemed to come as much from the fact that there hadn't been a general conflict in so long that karma dictated that at some point there simply *must* be a war. Still, while tensions between European countries grew, few ordinary Europeans truly believed a large war was on the immediate horizon. In *Dance of Furies*, Michael Nieberg details how, rather, most Europeans took comfort in the fact that the Great Powers did compromise when crises arose and that they worked to maintain peace. This faith continued in 1914, and the onset of the war was truly a surprise to the ordinary person living in Europe, who expected yet another compromise would present itself at the last minute.

The Start of War

On June 28, 1914, Archduke Franz Ferdinand, heir to the throne of the Austro-Hungarian Empire, toured the city of Sarejevo, the capital of Bosnia-Herzegovina. Military maneuvers had been planned for his inspection outside of the city. This region had recently come under the administration of the Austro-Hungarian Empire despite the fact that many in it and in the region of the Balkans opposed the reach of the empire. A strike against the Austro-Hungarians was planned by a group of young pro-Serbian men who saw the policy of the Austro-Hungarian Empire as diametrically opposed to the existence of a strong, independent Serbia in the region. One of these young men, Gavrilo Princip, succeeded in shooting and killing Franz Ferdinand and his wife on this day.

The date, June 28, was a symbolic one for the Serbian people. It commemorated the Battle of Kosovo, fought in 1389, which saw the Serbians defeated by an encroaching Ottoman Empire. The battle, though, in many ways signified the beginning of the Serbians' sense of national identity, and having the symbol of another encroaching empire in their backyard on this date seemed karmic.

There was great surprise at the assassination in the Austro-Hungarian Empire, though, perhaps, no real sorrow, as Ferdinand was not a popular person. But the government had to find out who was to blame for the assassination and who to punish. Serbia was at the top of the list because of a long-standing hatred by Austro-Hungarian officials of Serbia. These officials had long sought a war against Serbia. Serbia had been a thorn in the empire's side as the empire sought to expand its reach through the Balkans. Subsequent investigations led to blaming the Serbian government for the assassination, but this was not a trumped-up charge. Specifically, Serbia was found to have been harboring a terrorist organization that planned the assassination. More recent research has confirmed that Serbian government officials did assist in the planning of the attack and even supplied the weapons, though it also seems that the Serbian officials did not believe the plan would work.[1]

Still, this was an issue between Serbia and Austro-Hungary to resolve at this point. It was a topic of conversation for Europeans that summer, but it cast no dark specter over events. People still took vacations, traveled, went to work, and made plans for the future. It was only in certain diplomatic circles that, as the weeks went by, something more serious loomed.

The diplomatic communiqués flew around the European capitals in the days of July as Sean McMeekin eruditely describes in his book *July 1914*. Austro-Hungary sent an ultimatum to Serbia that stated that the only way to avoid war was essentially for Serbia to give up its independence and allow Austro-Hungary oversight for a number of Serbian institutions. Although Serbia was willing to compromise on a number of issues, full adherence to this ultimatum was not something Serbia was prepared to do, so the crisis continued.

As Austro-Hungary prepared for war against Serbia, other countries stepped in to the argument to complicate matters. Russia did not want to see Serbia subsumed into the Austro-Hungarian Empire, and looked as if she might enter into the conflict. Austro-Hungary, who feared a war alone against Russia, turned to its historic ally of Germany for additional backing as preparations for war progressed in July of 1914.

The German foreign minister sent a message to Vienna, essentially giving Austro-Hungary a "blank check" in how it chose to proceed. Notably, there was little discussion over the wording, and the Kaiser himself was not aware at what this message "promised." Although probably not intending to appear as giving a blanket endorsement of war, nevertheless Germany came to be seen as the aggressor nation that fanned the flames instead of one that imposed limitations on Austro-Hungarian dreams of vengeance. In the aftermath of the war, this decision is one of the elements used to peg Germany as being responsible for enlarging the conflict, and the "war guilt" clause of the Treaty of Versailles would confirm this international opinion. The punishment visited on Germany for this would be severe.

Focus now shifted to German military machinations. The German military had long believed that another war was likely, and when it happened, it would feature Germany fighting on two fronts in such a conflict—against Russia on one side and France on the other. Germany had long seen itself encircled by enemies. Therefore, military planners embraced a plan that would enable Germany to have success on two fronts. The plan involved a quick mobilization of forces that would head west, sweeping through Belgium and northern France to capture Paris. A tight timeline was necessary for this to work. Paris must be taken before Russia got her troops mobilized and in the field. The Germans expected that the Russian military would take at least six weeks to become mobilized. Russia managed to get mobilized in three weeks. Russian troops were poorly outfitted and did not perform well, but the German Army had to respond to the Russian threat and pull units from its western front to send to its eastern front ahead of schedule. And, in fact, others had got the jump on Germany with Russia mobilizing on July 31, France mobilizing (due to its treaty with Russia) on August 1, and Germany only ordering its own mobilization on August 2. The treaties and alliances that had helped to broker peace for the past decades now had the countries line up for war.

Additionally, the Germany Army had counted on an easy passage through the country of Belgium preceding the attack on France. This assumption proved faulty. The Belgium government, led by its king, realized they could not simply allow Germany to pass through its borders without in essence giving up their independence. Though with a significantly smaller contingent of troops, Belgium decided to fight Germany as German troops crossed the border. This slowed down the German advance as Germany now found itself unable to cross bridges that the Belgians blew up, and needing to expend artillery and soldiers to lay siege to and take Belgian cities. It proved to be a frustrating time for the German soldiers, who thought this part of the plan was supposed to be easy, and it led to numerous retaliations against Belgian civilians. This further cemented the viewpoint prevalent in Allied nations that Germany was behaving abominably.

As some German units made it into France that fall of 1914, French civilians saw and heard the war getting ever closer. Taxis in Paris were used to funnel reserve troops to the front. People in cafés in Paris could hear the artillery booming in the distance. At first it looked as if Paris might fall. But it was saved at the Battle of the Marne and continued to be safe for the remainder of the war.

These early weeks of war saw Allied (French, British, Belgium) armies in Western Europe moving significant distances to counter the German advance. The Germans too were marching over 150 miles in a short amount of time. At this time it was a war of movement. But in early fall, that all changed. German troops, after retreating from the Battle of the Marne, began to dig in, to dig their trenches. And so began that which is most associated with World War I—trench warfare.

The first trenches were hastily dug and not very deep. But they did offer a measure of earthen protection from the enemy artillery blasts, the sharp shooters, and the machine guns. The benefits of the trenches were clear. It was not long until both sides began building an extensive network of trenches to house their hundreds of thousands of troops. The German trenches were deeper, more organized, as if they meant from the beginning to be there for a substantial length of time.

The fighting that did result on the Western Front through Belgium and France shaped much of the world's opinion on the war. Millions of troops from both the Allied and the Central Powers sides fought one another over a comparatively narrow strip of land. Due to the high concentration of soldiers and weaponry in this space, it took on a moonscape-like quality over the months. Trees were wiped off this fertile land, as were villages, and it became unrecognizable. The soldiers found themselves living in these new massive networks of trenches, which offered both protection and horror. Rains turned the trenches and countryside into dangerous mud; the confines of the trenches offered no easy routes of retreat; and to go on the attack meant climbing out of these trenches and exposing oneself to enemy machine guns before taking even one step of advancement against the enemy.

Battles took place again and again over nearly the same spot of land—the First Battle of Ypres, the Second Battle of Ypres, the Third Battle of Ypres, the Fourth Battle of Ypres. Throughout 1915, each side hoped for a breakthrough. The year 1916 saw months-long battles take place in a concentrated attempt to eat up the opposing army. The German Army attacked at Verdun, knowing that France would do anything in its power to save that historic place. Casualties for the French numbered over 337,000 by the time the battle was finished. The same number equally applied to the German side—something the Germans had not fully considered. The British attacked the Germans at the Somme that summer. Casualties for all sides involved totaled over 1 million.

And the front still did not move. And the war continued to drag on. Different commanders were appointed on both sides. Each government hoped that the new appointment would prove the masterstroke and that, if not a breakthrough, at least a turn of fortune would come their way.

US Entrance into the War

As 1916 wound down, one lingering question hovering over the combat was if the United States would join the Allies and formally enter the conflict. US citizens had embraced the neutrality stance of the country since 1914. People in the United States were shocked by the casualty numbers from the war that they read about in their newspapers, and they congratulated themselves on not being a part of the murderous chaos going on in Europe. This was a presidential election year in the United States, and Woodrow Wilson campaigned for reelection while his supporters touted his ability to keep the nation out of war. Wilson himself never said such, as he was no hypocrite on this issue at the time. Wilson had a number of negative traits and perhaps approached this subject more with an eye toward the public outrage that could come if he made such statements and then pivoted; nevertheless, he remained consistent on this issue in these months. He had already realized the great likelihood that the United States would find itself enmeshed in the war soon, not because it would be drawn into the war, but rather because the moralistic Wilson saw an opportunity to try to craft the peace that would follow. The only way to gain an invitation to a peace conference, however, was to be a participant in the conflict. In 1916, he already knew the war would not be far away for the United States.

This likelihood of war would prove true only a few weeks after the election. In January, Germany announced a new policy—unrestricted submarine warfare. This was a calculated risk on the part of Germany. Since 1914, the United States had claimed that as a neutral nation, it should be free to trade with whomever it wished. This included sailing in waters that were war zones as, obviously, a noncombatant. When American cargo or lives were lost because of German torpedoes, Wilson demanded apologies from the German government, which he got. The United States continued to sell supplies and make loans mostly to the Allied Powers rather than the Central Powers. The bias was already clear.

Unrestricted submarine warfare would target any ship sailing in a war zone, whether it flew the flag of a neutral nation or not. Germany knew this policy would enrage Wilson and offend the United States to such an extent that the United States might enter the war. But Germany also calculated that the United States would enter at some point anyway and that it was better for it to happen soon so that Germany's victory over all of its enemies would be forthcoming. Germany was also gambling that it could extract a victory over Britain with this submarine policy before the United States would even arrive in Europe. Much of Britain's supplies (and money) were coming across the Atlantic, and Germany meant to strangle the cargo bound for Britain and force her to surrender. However, the convoy system that the British implemented gave protection to her ships even before the United States would be fully engaged in the war. Then, on March 1, 1917, US newspapers released transcripts of a German telegram to the government of Mexico in Mexico City. In the Zimmerman Telegram, as it was known, Germany offered an alliance with Mexico. If Mexico was willing to engage the United States on its northern border, thereby occupying US troops who would not be available for the Western Front, then Germany, on defeating the United States, would grant to Mexico large areas of the American southwest, including Texas, New Mexico, and Arizona. If the vague threat of unrestricted submarine warfare had not pushed some Americans over to the pro-war side, the Zimmerman Telegram certainly did. Indignation was great, and there was overwhelming support for a war resolution that was approved in the first week of April. The United States was joining the Allied effort. Troops would first reach Europe in July 1917, but the bulk of the large fighting force would not be in place for use until 1918.

The entrance of the United States into the war gave a boost to the morale on the Allied Western Front, which had suffered mightily over the past years. Fresh troops by the thousands arrived, and with them new supplies. The working relationship between US military officers and the French and British officers was a work in progress. They did not always agree on what the next steps should be. For example, the British expected that US troops would fill the holes in their own mangled regiments, whereas the head of the American Expeditionary Force, General John J. Pershing, refused to have US units broken up in such a way.

In 1918, other events away from the Western Front were having an influence on it, however. The Russian Revolution, led by Vladimir Lenin, began in 1917, and once in power, Lenin did not want the distraction of war for his country. In fact, extracting Russia from the war had been one of his selling points of the revolution, with its slogan of "Peace, Land, and Bread." The Russian withdrawal from the Allied side in 1918, which involved negotiating a separate peace treaty with Germany, meant that Germany was now free to commit more of its troops on the western side of the war. But Germany in 1918 was not the same Germany as in 1914. Despite the continued military efforts to achieve victory, the war was taking its own toll on the home front. There were shortages of food and fuel, talk of mutiny among sailors, frustration among civilians, and ultimately, in late 1918, this strife would lead to a governmental overthrow of the monarchy. The kaiser went into exile. Germany's allies of Austro-Hungary and the Ottoman Empire had also suffered severe setbacks that rendered their military contributions in the last months of the war moot.

On the Western Front, the Allies planned an offensive for the fall of 1918, hoping that with the new US troops maybe that often ached-for breakthrough could finally occur. The Meuse-Argonne Offensive began in September 1918. The United States launched over a million of its soldiers at German troop lines and supply lines. Allied troops managed to cut off German supplies to the front, and with the social disruption in Germany, the country's new political leaders decided to ask for an armistice. The Allies and the Germans reached an agreement for a cease-fire, and the armistice began at 11:11 A.M. on November 11, 1918. The Germans believed the basis for the peace would be the Fourteen Points Plan that President Woodrow Wilson had announced in January 1918. This was a plan designed to create a new world order of sorts that respected the self-determination of nations, rights of neutral nations, and a new collective organization called the League of Nations that would provide a space for civil discourse for all countries. Peace negotiations would formally begin a few weeks after the armistice and carry on for six months before the Treaty of Versailles was signed in 1919.

The final treaties (there were several signed) highlighted not a sense of magnanimity, however, but rather a sense of pent-up revenge. The Allied nations of France and Britain blamed Germany for the war and wanted compensation to be paid. Wilson suffered health ailments while at the negotiations, and his efforts to redirect the treaty away from full-out punishment of Germany failed. The war reparations assessed on Germany would spell the total collapse of its already fragile economy. Germany was forced to accept a treaty that not only blamed Germany for the war but also instituted billions of dollars in payment that would be due the victors. Hyperinflation would hit the country by 1923 as the new Weimar Republic government in Germany tried to cope with the reparation payments, and paper money would become worthless as the government went on a printing binge to fulfill their payments.

The political and economic instability that followed would give rise to right-wing groups like the National Socialists, who would dominate the government within a decade. These events and decisions paved the way for World War II.

This war rewrote European and world history. The size of it, the sheer number of casualties, the brutality of the weapons all changed calculations about any future war. Fighting this war from an administrative standpoint, a foreign policy standpoint, or a front-line solider standpoint took massive adjustments to people's expectations about war. Four empires crumbled in this war, and the men who fought it saw brutality and inhumanity on display for years, with seemingly no end. This inhumanity gave rise to a sense that civilization itself was collapsing. The old standards of decency and behavior were gone. The world of the 1920s was vastly different, with societies having lost their innocence and optimism for the future. The poetry and literature of the 1920s saw writers grapple with this loss, trying to make sense of the insensible.

Logistics of War

The number of soldiers who served in this war was equally staggering. Among the Central Powers of Austria-Hungary, Germany, the Ottoman Empire, and Bulgaria, nearly 23 million troops served. The Allied countries (of which there were many) mustered over 42 million troops, with the French contributing over 8 million, the British Empire nearly 9 million, and the United States over 4.5 million. Russia put 12 million in uniform, and Italy over 5 million.[2] These soldiers suffered casualties too, like no armies had in the past. The Central Powers had over 8 million wounded, and the Allied Powers close to 13 million.

Total deaths of soldiers at the end of the war stood at approximately 8 million. The number of civilians who died was similarly mind numbing, at another 6.5 million. These noncombatant deaths, though, do not include deaths due to the flu outbreak in 1918, which most certainly turned more deadly because of the mass movement of people associated with the war. As Michael Clodfelter points out in his massive volumes on war casualties, if one factors out the deaths on the Russian front (which was a unique and horror-filled front) in World War II from the Second World War totals, the total of military deaths from the rest of that war would be only half that of World War I.[3]

With numbers such as these, caring for and treating the millions of soldiers meant medical care on an unprecedented scale for these armies. Medical care, just as the war itself, would need to be industrialized, streamlined, and optimized. It had to be efficiently run and administered. And more than that, it had to work. The soldiers depended on this care, obviously, but so did the governments and bureaucracies that knew the medical departments must be given the support to heal the wounds of these men as often as was humanly possible.

And simply fighting and surviving in this war for the millions of soldiers who were injured in the war did not mean returning home in the same physical or psychological state as when they went off to war. In the case of the American Civil War, the largest more recent war in terms of manpower, tens of thousands survived the war, but with some injury as many photo-journalists detailed in the years after 1865. Although injuries might be of any sort, the number of soldiers who received amputations was pronounced. It was not uncommon for soldiers to lose more than one limb and survive. Survival, while certainly a positive outcome, often also meant that families had to cope with a member who might no longer be able to hold the job he once had. He would need care for the remainder of his life. As a veteran, his government also had a continuing duty to assist in that medical care. The Allied governments were well aware that the responsibility for care would go on after this war ended.

It was paramount in World War I that troops be returned to the line if possible following an injury. But it also was important that if they could not fight again, they be given the best medical care they could, considering the risks and sacrifice they undertook for their countries. Therefore, as these armies moved, so too did the medical front of the war. Thousands of doctors, nurses, and other medical staff worked to provide the best treatment they could in overwhelming conditions. Because of their work and organization, an equally stunning number of men arrived back home after World War I, having survived injuries that previously would have proven deadly.

Medical Advancements Prior to World War I

The ensuing years between these two large-scale conflicts mentioned, the 1860s to the 1910s, saw a number of medical advances that would substantially aid the efforts of the medical personnel in World War I. Perhaps chief among these was the use of disinfectants in hospitals and surgeries and the understanding of the germ theory of disease. Joseph Lister, a British surgeon, pioneered the use of sterilization in hospitals through the use of carbolic acid in the 1860s and 1870s. He required the staff handling the surgical equipment to wash it in carbolic acid to kill the germs. Dressings soaked in carbolic acid were used on wounds, to stem infections. The use of such antiseptic practices was quite controversial, though, as it violated the traditions in many hospitals. Many doctors contended that such measures were unnecessary, but the effect of their use was undeniable. Infection rates plummeted in facilities that adopted these measures.

Obviously, progress was afoot. As the next generation of doctors and nurses were trained, such protocols became more standardized, and patients benefited from lessening the risk of infection. Technological innovations would also begin to have an impact on medical matters. Inventions such as the X-ray machine would go from a circus trick to a regular part of diagnostic work in medicine.

Surgeons made numerous advances in their chosen specialization. Brain surgeries were being conducted, and surgeons had moved to wearing rubber gloves and having as aseptic an operating environment as possible; surgeons such as George Crile were studying shock. The studies into shock moved Crile to begin tilting his operating tables so there would be decreased blood flow in the part of the body on which he was operating, thereby minimizing the risk of shock. Monitors for blood pressure began to make their appearance in operating rooms as well.[4]

The medical staff and patients in the Great War would benefit immeasurably from these types of recent scientific discoveries. Physicians had become quite adept at diagnosis, but they were moving into a world in which they could now offer medical therapies to truly help their patients.[5] However, chronologically the war felt short of providing a number of advances that have reshaped medical practices. Of these, antibiotics such as penicillin (discovered in 1928, 10 years after the war's end) have made major differences in the medical treatment offered in the wars since. The great panoply of pharmaceuticals that would be turned out of laboratories after the 1930s would end up reshaping World War II in a way that the doctors and nurses of the Great War could not have dreamed of.

Therefore, on the medical front World War I also is as much a bridge from older modes to modern warfare as it was on the military technology front. Medical personnel were confronting a modern industrial war without all the medical know-how of the rest of the modern twentieth century. Although the pharmaceutical side of treatment was beyond what was available to them at the time, new techniques, new understandings of healing, and the importance of immediate care all made a difference on the medical front of this war and improved the outcomes of the patients over past armed conflicts.

Technology of the War

It is arguable that World War I is not truly the first worldwide war. The Seven Years War from 1756 to 1763 had soldiers from North America, Europe, and Asia fighting in it in locales in all of those continents as well. However, World War I is the first war that had the massive movement of troops, the enormous scope of carnage, and the redrawing of boundaries that seems befitting of its moniker. It saw empires dissolved and expanded by the time of its end. It also introduced into modern political rhetoric such ideas as the self-determination of nations and, in President Woodrow Wilson's words, the idea that the world should be safe for democracy.

Many historians have written about the technological innovations of the war that enabled such carnage to occur. Military weaponry changed significantly in the late 1800s, particularly with industrialization. This meant change in the properties of the weapons themselves and in the production ability for such weapons. Weapons had become more automated with the use of repeating rifles; the development of the Gatling gun, which was one of the first rapid-fire guns; and the desire to continue to increase the rate of speed with which guns worked. The Maxim gun, designed by Hiram Maxim, appeared in the 1880s and was the first true machine gun. There was no turning back. Industrialization ensured that these guns could be fabricated and produced in large quantities. The use of machine guns meant that more soldiers would suffer not only bullet wounds but wounds made with bullets traveling at a higher rate of speed than years earlier. Bullets themselves (a pointed projectile), instead of round shot, were new for this war. The industrial capacity of these countries delivered piles of weapons and ammunition to the front lines as had never been seen before. The need to deliver these weapons and their ammunition meant that governments had to take on the responsibility of making sure there were enough workers in munitions factories and that government contracts were adhered to. It required substantial government oversight of private industry.

In addition to the weapons that were handled by single individuals, the war saw larger and larger pieces of artillery be built, including the massive siege guns that the Germans used in northern France. The "Big Bertha," as it was known, was a 420 mm howitzer gun produced by Krupp in Germany. It had to be transported by rail and tractor, and weighed in at around 47 tons. It was the largest such gun on Earth and played a significant role in fighting in Belgium at such fortress towns as Liege. Even these large weapons, though, became unnecessary by 1916—they were not big enough to take the walled city of Verdun, nor could they outdistance the new artillery that the Allies had developed. Krupp then built the rightly described colossal Paris Gun for use in the last couple of years of the war. Its barrel alone weighed over 130 tons and could target areas 75 miles away. It had the longest barrel of any artillery to date. Though this particular weapon did not play a decisive role in the war, the psychological intimidation such weapons could cause was unparalleled. It was almost as if manufacturing hit up against the physical limits of deployment by this point, however.

The British Royal Navy pioneered the use of the landship (because it was part of the navy), or what would become the tank, at the Battle of the Somme in 1916. This was the brainchild of Ernest Swinton, who saw the usefulness and flexibility such a weapon could bring to moving infantry formations (though others were also simultaneously working on such a project). His attempts to sell the British Army on his design went nowhere, but the British Navy, with Winston Churchill serving as the First Lord of the Admiralty, was interested. Painted battleship gray, the landship would bring heavy artillery physically onto the front lines instead of it being in a fixed position in the rear of an advance. These early tanks promised perhaps more than they delivered. They were prone to breakdowns and moved slowly when they did move. The mud that hampered all movement on the Western Front debilitated these machines as well. But they presented a tantalizing view into the more modern warfare of the twentieth century.

Aerial combat too found its birth above the countryside and towns of Europe. In the few short years of the airplane's existence, huge strides had been made in the design and airworthiness of the contraptions. The French have always had a great interest in aviation. By 1914, some two-man planes with 80 horsepower engines were fielded by the French. Airplanes offered a unique opportunity for all sides to obtain reconnaissance on their enemies, seeing from above where armies were moving and where the flank might be. Airplanes continued to fill this role but also took on operational duties as pilots engaged the enemy in battle, and they began to be utilized as low-altitude bombers. Planes, fitted with bombs and guns (including handguns carried by the pilots), were thrown into action above the European battlefields. In many ways these aerial fights continued the romance of war and heroism that became defeated and absent on land. One man could display bravery in the air, and that bravery could turn the tide to victory. Such romantic notions had no place in the trenches and mud below.

Ordinary people followed the stories of the flying aces and how many of the enemy they shot down. Planes such as the German Fokker and the British Sopwith Camel became recognizable in newspaper photographs. Other types of aerial craft also became increasingly known—Gotha bombers and the zeppelin airships. Both were launched by the Germans and bombed positions not only in France but in Britain as well, with London being a prime target. Hundreds of civilians were killed, and thousands were wounded. There was also substantial property damage. Although in whole numbers, these losses were far below what would be seen in World War II; the loss of civilians due to what would be random chance also affected people's opinions of how this war was like none other.

The modern war practice of targeting civilians is clearly in evidence with these bombing campaigns.

The prevalence of the German U-boats, the submarines, also provided the start of this now common aspect of naval warfare. Other countries had experimented with the idea of a submarine in past decades, but the Germans made it an important part of their naval arsenal. In the days before radar, the U-boats operated in a great deal of secrecy. They patrolled the war zones around Europe, hoping to torpedo supply ships funneling equipment from North America to their enemies. They also would be used against troop ships arriving for combat in Europe.

Aside from simply building larger versions of already-used weaponry, new weaponry also made an appearance in the war. Items such as flamethrowers became used in battle. This could set fire to a section of trench quickly and was fearsome when used against groups of men. The Germans weaponized poison gas and began using it in the first part of the war. Trial runs were made before the large-scale use at the Second Battle of Ypres in 1915. The Allies were soon to follow. The German chemical industry had made a number of advancements in the period before 1914. One advancement was the sheer size of the industry. In the early twentieth century, most of the world's chemical dyes came out of German factories. There had also been developments with various types of acids and inorganic materials, including chlorine. German scientists had pioneered work with industrial chemicals, resulting in some of the modern fertilizers that would be used by farmers across the world in the twentieth century. Scientists from other countries would travel to Germany to study and learn from these German industrial pioneers. There was a great deal of worldwide respect for the German chemical industry until World War I.

Germany decided to use its industrial advantages in all ways when the war broke out, and this included using shells and chemicals to release poison gas in World War I. They were the first to do so, but they would not be the last. In making this choice, they violated international agreements that banned the use of asphyxiating gases. Germany had signed on to these treaties twice—in 1899 and 1907. They used tear gas and chlorine gas, and by the end of 1915, they had progressed to using phosgene gas, whose shells were marked with green crosses.[6]

Despite the fact that poison gas such, as phosgene and chlorine gas, was used by both sides, the Allies never achieved the efficiency that the Germans did in gas production. In 1915, gas masks became de rigueur accoutrements for all soldiers in the front lines. The gas masks themselves would undergo various design changes throughout the war to improve their filtration systems and limit exposure.

Other countries then built on this to develop their own versions of chemical weapons. The British, and later the United States, would deploy their own gas weaponry in the war. Although today the use of gas is seen as a moral and ethical problem, there were no such major qualms in World War I. If one side had it, the other side sought to get it. Not to deploy it because of concerns about morality was not really debated at the time.

All of this technology brought the war home in an immediate sense for those civilians near the battlefields too. Towns and villages near the front lines were eventually abandoned as the war progressed. Artillery barrages destroyed churches, schools, homes, and farmland up and down the Western Front. People became displaced by the hundreds and then the thousands.

The use of such technology reshaped what warfare meant both for political leaders and for the front-line soldiers. Learning to live with the constant artillery shelling or the fear of a gas attack affected soldiers psychologically in ways that would be studied after the war. It took commanders a long time to realize how many of the hallmarks of battle in the nineteenth century—cavalry charges, large breakthroughs with infantry—were simply unworkable on the mechanical battlefield of the twentieth century. This adjustment would also have to be made by the troops. The hero's dash across the battlefield simply couldn't happen under a withering machine gun fire. Famously, the soldiers adapted more quickly than the headquarters staff. The soldiers saw the futility of such a dash early on, but Allied headquarters kept ordering charges "over the top" and out of the trenches through the majority of the war. The charge that the British Army's soldiers were "lions led by donkeys" did have some semblance in fact. Some British and French officers continued to try for a breakthrough long into the war. Other officers adopted new tactics for this new warfare and were more responsive to the conditions of the front lines.

In some ways the technology benefited Germany, as it was the most heavily industrialized power in continental Europe, and it was able to use that to its advantage for much of the conflict. Germany also occupied industrial and coal-rich regions in Belgium and France once the war began. Throughout the conflict, Germany continued to turn out weapons, be they ammunition, artillery, submarines, or chemical weapons, at a relentless pace.

This didn't mean, though, the Allies couldn't compensate. The combined armies of the British and French were able to match the German one. And the Allied home front attacked the manufacturing needs of the war successfully. The government in Britain helped organize the efforts of turning out munitions. Thousands of female workers took up jobs in the munitions factories to do their part to provide for the war. Investments from the United States helped support the Allied spending, and then US entrance into the war gave a much-needed boost that helped topple the German government's commitment to the war.

Soldiers in the War

For ordinary soldiers on the Western Front, the war turned out to be not what they expected in 1914. Previous wars had often been fought with professional soldiers, but because of its scope, ordinary civilians would need to become soldiers for this war. Civilians joined through volunteering and through national drafts that were instituted by the governments. When the war began, men across Britain and France (and Germany) excitedly volunteered for what promised to be a short war, and then found themselves mired in the trenches for months. The widespread use of trench warfare on the Western Front set it apart from the more mobile Eastern Front. The soldiers found the conditions near inhuman at times. Soldiers were constantly exposed to the elements. The cold could be mind numbing and the rain relentless. The mud was pervasive. In some areas, the intense shelling meant that soldiers could not be relieved from forward positions. They had to stay there if they were wounded or even killed. There were stories of rats and other vermin that would crawl into the trenches to feast on the unevacuated dead. And that if a soldier remained still for too long, he might start to feel a nibble himself. For the troops being forced to stay in such situations for days meant more than simply being uncomfortable. It could be a shattering experience. It compromised their physical health, and it was a severe psychological strain even for those who emerged physically intact from it.

Soldiers would describe never feeling warm in the winter months in the trenches. Being constantly exposed to the elements could lead to hypothermia. And the trenches themselves with their moisture could cause other ailments. Trench foot was a common problem. Soldiers found themselves in water or mud so often, their feet in their leather boots never would fully dry. Sores would develop, and skin would slough off. The pain was debilitating. It also meant that the area was ripe for infection. And bacteria were everywhere. Trench fever was spread by lice from man to man, and it was difficult to eradicate.[7]

Some months saw rain fall every day. It rained in the spring. It rained in the summer. It rained as the summer turned to fall. The mud became omnipresent and unbearable. Horses went down in it. Men went down in it. Duckboards of wooden planks were strewn across the worst parts of the land to keep people out of the muck. People thought of it as a quicksand. Once someone got grabbed by it, it was difficult to pull him out. Some survived hours with mud up to their chests, but once rescued, they succumbed to the internal injuries of all that pressure having borne down on them. Drowning in the mud seemed morbidly symbolic of this war.

One part of the year when a soldier didn't have to worry about mud was in the winter. The experience in winter, though, could be otherworldly. One soldier later wrote:

> It was bitterly cold and all the ground was frozen hard... had socks pulled over our hands in place of gloves. ... It was very clear weather and every sound carried, so that we moved carefully and slowly. The main trench was a long black-shrouded ditch full of dark figures, scuffling, muttering to each other and there were hissed curses when a steel helmet clanged against a rifle.[8]

Frostbite was a real concern. Exposure was a given with men out in the trenches in the winter. Unlike in past wars, the soldiers of the Great War did not go into winter quarters. Instead, they stayed in their positions through the bitter months.

Soldiers spent their days secretly peering at the enemy, doing repair work on the trenches, and simply trying to keep their heads down. Nighttime raids were ordered occasionally to check on the defenses in no-man's-land or to gain reconnaissance information either visually or even by kidnapping an enemy soldier. This was the day-to-day activity of being in the frontline positions. But there was still always the threat of enemy artillery and shelling that occurred near daily, causing not only disruption to the day's plans but also numerous injuries for those in the trenches. Enduring the noise and concussion was difficult. It could be "loud enough to split the eardrums" and cause hearing loss.[9] The smells of the front line were intense. "The stench of the battlefield was a mixture of decomposing bodies, the chloride of lime spread to combat infection, creosote used to deter flies, human and animal excrement, smoke from spirit stoves, and human sweat."[10]

Injuries from combat were numerous, but other types of injuries would occur too. In such circumstances as war, accidents were common. Guns or artillery might be misloaded and explode. Friendly fire could rain down as artillery crews found their range. There might be vehicle accidents, or trains would derail, causing injuries to those on board. Given the large numbers of horses that played a part in World War I, there were even accidents in corrals and in the training of the horses. Many of these injuries would resemble what might happen in the civilian world while others were distinctly a product of the war.

The burden of death and injury was omnipresent. Soldiers faced daily reminders of death, even if they were not taking risks themselves. Will R. Bird of Canada vividly described two such experiences in subsequent days:

> We had got used to the slamming roar of gun fire, and now we heard machine guns barking and snapping, and bullets come singing overhead to go swishing in to the distant darkness. Some struck on wire or other obstructions and we heard the sibilant whine of ricochets. We had sandbags to fill. One man held them and the other shoveled in the gruel-like mud. When twenty or more were done, a man jumped up on top and emptied the bags as they were handed up to him. It was ticklish work and one often had to jump into the trench as bullets were humming about all night. …

> ... this new smell halted us. A corporal stepped from the gloom. "Here's bags," he said. "Go in there and gather up all you can find, then we'll bury it back of the trench. Get a move on." A flying pig had exploded as it left the gun and three men had been shredded to fragments. We were to pick up legs and bits of flesh from underfoot and from the muddy walls, place all in the bags and then bury them in one grave. It was a harsh breaking in. We did not speak as we worked.[11]

What many of the recollections of these soldiers reflect was the role of uncertainty in being on the Western Front. The soldiers never knew when a stray shot might hit them or when they would be summoned to move forward or march. Innocent maneuvers might go terribly bad, and there often seemed to be a sense of morbid curiosity in the texts—why them and not me? Was it all just random luck? There was no way to make sense of it all because it simply didn't make sense. Many diarists commented that even after returning home and having years pass, they still did not understand everything they had gone through. Some hoped to exorcise the past by writing about it. Others chose to bury the past and not speak of it.

Being under fire was an experience like no other. During periods of combat and shelling, soldiers received bullet and shrapnel wounds by the thousands. Those who were wounded in the no-man's-land between the two opposing armies would be evacuated at night or in some cases only after days of intense shelling. Shells would land in the trenches, destroying entire sections at a time. The troops found themselves being ordered into ghastly attacks that stood little chance of success. The morale of an army could hardly be maintained through such conditions when these types of attacks occurred. Yet the armies proved to be amazingly resilient through the years.

The scale of this war was immense. So too would need to be the medical organizations to support the troops. The conditions of this war were unlike anything seen before, and those that hoped to treat the wounded would have to adapt to their new reality quickly. Engagements produced hundreds and then thousands of wounded soldiers who would have to be triaged and treated as quickly as possible. Trial and error proved to be the pattern in the early years as military medicine struggled to fully staff all the facilities it needed to. It also had to cope with the numerous volunteers who appeared in the frontline hospitals. This happened across France and Belgium as doctors and nurses appeared either by design or by choice. Trial and error also applied to the treatment itself of the patients. Although military doctors may have had exposure to the types of wounds inflicted by bullets and shrapnel, the large number of civilian practitioners who would be brought into this war had no such background. These doctors had been practicing in small towns or cities and treating the assortment of workplace injuries, normal illnesses, and accident cases just a few months earlier. They would need to get up to speed quickly, and the medical establishment on the home front played a role in disseminating information to medical personnel there about the types of wounds and treatments that were being seen and used on the front lines.

The high number of patients that would come through their beds meant there would be ample opportunity for medical personnel to see what treatment plans worked best. Doctors had an ability to experiment and innovate during the war, developing new techniques and procedures. "The war has produced a mental quickening which has advanced our knowledge of the mechanical and the scientific far beyond that of a like period of peace."[12] They also had the ability to institutionalize the use of medical equipment such as the X-ray machine. In the case of the X-ray machine, it had been around for approximately 20 years and had been used sporadically with military medicine, but this war changed protocol in how it was used and how it would be used in future conflicts.

Often lost in the shuffle at the end of the war was the true nature of the role of medicine in the upheaval. The ability to provide quick and innovative treatment became more expected of military medicine, which certainly created a better environment for armies in future conflicts. The Great War seemed incredibly pointless at many times during the fighting and as people later looked back on it. The role of the medical front, though, demonstrates that some good came out of the war. Though one must be clear, this viewpoint would feel blasphemous to the contemporaries of 1914 to 1918. The good that arose from the medical front does not sanction the idea that war was "worth it" for these new medical understandings. But it was not naive to take note of these new elements and realize that people in the future could benefit from them.

This focus on medical "news" was not simply for scientific know-how of the medical community, however. It was also imperative to successfully treat the wounded, as these men would be needed, if at all possible, to fight another day. Soldiers simply had to be back in the line, if they could be, to keep the armies sustained. The troop strength needed on the front dictated this. If they could not be returned to their units and needed further care, that care was the responsibility of the country's government because these men had put themselves at risk for this war. So the wounded became a part of the home front medical world too. Patients had to be "processed" through the medical front expeditiously, just as they had to be "processed" into the military and deployed. Having a successful medical outcome for the wounded was needed by the militaries, the governments, and the individual soldier.

2

Allied Medical Innovations

As the war dragged on for four long years, hundreds of thousands of soldiers among the Allied Armies were given medical care. With the slew of patients, medical innovations and advancements during World War I occurred continuously throughout the conflict. World War I is often cited as the dividing line for the beginning of the "modern" world. In the case of surgery, this is an applicable statement. The doctors in the war greatly benefited from the knowledge gleaned in the past decades that explained germs and bacteriology and from inventions such as the X-ray machine, but the war made what are now recognized as modern surgical techniques and treatments to be put into place uniformly in the treatment of patients.

Improved treatment came in many areas. Treatment obviously was affected by new surgical and medical techniques. However, transportation, availability of doctors, and the organization of the medical services all had bearing on the improved treatment offered the wounded. Medical specializations saw more differentiation and new therapy methods. The care of physical wounds to the body saw significant advancements that lowered infection rates. Aside from the immediate aid to soldiers, these advances in medical treatment benefited civilian medicine and surgery after the war ended. Major George A. Stewart of the War Demonstration Hospital of the Rockefeller Institute for Medical Research believed that the war had "advanced our science half a century in four years."[1]

Roger Cooter and Steve Sturdy write in the introduction to *War, Medicine, and Modernity* that history should deal with "the reciprocal processes of the civilianization of medicine in war and its militarization during peacetime."[2] They point out World War I did not allow any separation between military and civilian medicine. As they noted, the Great War provided "a crucial site for the development of new kinds of medical organization and labour."[3] It is important to stress the development of strong specializations and the bureaucracy of medicine because the effects of these elements are too often neglected in examinations of the subject.

The term *triage* originated during World War I. It came from the French word *trier*, which means "to sift." Wounded soldiers were "sifted" into one of three categories to expedite their treatment. If nothing could be done for them, they were classified as "expectant." If immediate surgery would aid them, they were classified as "immediate." If their treatment could be safely delayed, they were classified as "delayed."[4] Interestingly, this type of categorization was very much a product of this period of progressivism, industrialization, and modernization. The focus on efficiency that was a hallmark of the early twentieth century transferred easily to at least this arena in the war. As noted at the time, the British military had "practically the same general system of caring for their wounded as the French."[5]

Stretcher-bearers and field ambulances were utilized to remove the men from the front. This in itself could be dangerous work if a battle or continued shelling was still underway. The stretcher-bearers had to go into the area known as "no-man's-land" between the opponents to carry out the wounded. The injured soldiers would first be taken by the stretcher-bearers to the regimental aid post. The wounded that could walk in were also seen at this post first. If the injury was minor, the patient would be kept in the area and treated at an advanced dressing station. The patient would then be sent back to his unit. If the injury was more serious, the wounded were taken to casualty clearing stations near the front lines. The armies established field hospitals as far back as 8 miles from the front because of the range of artillery. Larger hospitals such as base hospitals could be found an additional 10 miles back. Transportation between these hospitals and the front lines was of paramount importance to ensure prompt treatment.

Proper surgical treatment was beholden to the transportation services that brought the wounded soldiers to the doctors. Obviously, the quicker the wounded arrived, the better, but if a wounded man did not arrive within 24 hours of receiving his injury, in some cases surgery would not even be attempted.[6] In a period of little battle activity and barring circumstances that separated a soldier from his trenches, wounded soldiers would arrive at an evacuation hospital between 4 and 24 hours after they received their injuries.[7] Ironically, the development of trench warfare allowed for "efficient work on the part of the medical department, in that the latter has time to perfect and maintain its organization."[8]

The delivery of patients was overseen by the Ambulance Corps. Ambulance Corps transportation to hospitals often suffered from organizational problems, and the different Allied armies had varying philosophies for their use. For example, the British Army had no less than 83 different types of transports for the wounded.[9] The wounded were removed most often on wooden carts pulled by mules early in the war. Mechanized transport via automobile became more familiar in the later stages of the war. These motorized ambulances varied in size themselves, carrying perhaps four or six stretchers filled with patients, or more than that if patients were ambulatory.[10]

There was further disagreement on exactly what purpose this transport was to serve. Was it merely relocation of the wounded? Or was it an opportunity to offer treatment as well as relocation? The French medical service did not dispense treatment while transporting the wounded. In contrast, the United States felt that transportation between first-aid stations and field hospitals should be more than mere rides. Consequently, the United States treated its transportation services as an extension of the medical services.[11] Although medical treatment was available in ambulance rides during wartime, it was not commonplace for civilian medical emergencies in the United States until the 1970s.

The descriptions of the Field Ambulance encompass more than simply transportation. The French and other countries in Europe used the term as a general one to signify hospitals, usually a mobile hospital.[12] Field ambulances operated advanced dressing stations that performed a number of tasks. Dressings might be reapplied at these stations; adjustments to bandages and "immobilizing apparatus" occurred; and there would be another assessment of how seriously a patient was wounded. Anyone who was wounded, even if the wound was deemed superficial, was given an anti-tetanus serum.[13]

Casualty clearing stations were designed to take the more serious cases, and became larger over the course of the war. The casualty clearing stations that early in the war were staffed with only six medical officers and eight orderlies often gradually developed into standard evacuation hospitals. Great Britain's Medical Corps saw a thousand casualties a day in some of her casualty clearing stations.[14] Doctors performed surgeries at these stations, and with the US Army there were separate casualty clearing stations to deal specifically with abdominal wounds.[15] These were often adjacent to rail lines, and patients could then be easily sent on further to the rear for more follow-up care. However, there was great variation in the facilities of such stations. Some were in fixed buildings; others were in tents. Here too doctors experimented as to the best methods of treatment for the patients that came to them.[16]

Ambulance trains then transported patients from the casualty clearing stations to base hospitals (also referred to as general hospitals) in the rear. There was rarely any specialization associated with the hospitals, as it would contribute to longer transport times if a hospital could not take any and all patients. These hospitals ranged in size throughout the war, with some early in the war having a 500-bed capacity and some constructed toward the end of the war having a 2,500-bed capacity.[17] Surgeons found that anesthesia of nitrous oxide and oxygen was the best choice for most cases, and they worked to refine apparatus for administration of anesthesia during the war. This eventually resulted in the development of Boyle's machine, which featured reducing valves and controls for additional agents that allowed for longer-acting anesthetics to be used in surgery. Publication of the details of the machine came at the end of the war.[18]

When patients were more stable, they could be dispatched to hospital ships. This made room for future patients, and in some cases, particularly those of British soldiers, this would facilitate their journey back to the United Kingdom, where they would be given further treatment. Once patients were recovered from their injuries, either when still in France or in England, they would be sent back to their units. There were some wounds such as severe orthopedic ones and facial ones that would take many weeks or months of treatment. If doctors determined the wounded soldier was no longer fit for service, he would be sent home.

Wounds

The process by which doctors tended the wounded became paramount. Different armies in the war suffered varying rates of infection, but the danger was ever present. Early treatment of the wounded was key because of complications from loss of blood, danger of infection, and gas gangrene being so high. Part of the reason for these infection fears on the Western Front stemmed from the fact that much of the front itself covered previously tilled farm fields. Farmers had been using various types of fertilizer for years, and these chemicals were still present in the dirt, wreaking havoc on open wounds.

During World War I, doctors and surgeons faced wounds that they had not seen previously. When the war began, Allied surgeons were truly products of their time. Old bullet injuries were characterized by a straight path and a relatively clean wound. First aid for this type of wound only consisted of placing some antiseptic solution (by the time of the war, tincture of iodine was most commonly used) and a clean dressing on the wound. Often nature was then free to take her course. The main effort was to find good antiseptics and apply them. Little else was done. As imaginable, infection rates were overwhelming when this was the only method of treatment. "The surgical outlook at the beginning of the War was responsible for disasters in all the armies, and a radical re-organization became necessary at the beginning of 1915. Surgeons were obliged to admit that all war wounds were infected, and that tetanus, gangrene, and septicaemia, followed many lesions of benign appearance."[19]

Pointed bullets, as opposed to flat or round-shaped tips, were first introduced in World War I. When these new pointed bullets hit a person, they inflicted far worse damage than the older models did. The center of gravity was set further back in them, and the bullet tended to tumble once it entered a person's body. This tumbling motion tore soft tissue apart and shattered bones.[20] Its impact was likened to that of an older exploding bullet.[21] Doctors quickly realized the change and sought to inform their colleagues of the different wounds this new ammunition caused. By the end of 1914, the *British Medical Journal* had published several articles describing the bullets used by various armies in the conflict and what doctors could expect in wounds caused by them. The German bullets possessed the sharpest point and therefore rotated the most on impact.[22]

Doctors decried the use of the "so-called humane rifle ball" used in the Great War. It was noted that bullets used even a few years earlier, such as in the Spanish American War in 1898, had a center of gravity "near the middle of the mass" and "made wounds, which were, comparatively speaking, quite humane" as physicians tried to grapple with the extent of the injuries they were now seeing. The newer bullets, fired from a distance of 500 yards or less, were found to have "explosive effects."[23]

Doctors found that artillery injuries were not as high impact because of the lower velocity of the projectile but did result in having more clothing and dirt forced into the wound.[24] It is unclear whether doctors felt differently as the war progressed and the size and power of the artillery used became greater.

With this sort of damage inflicted, cleaning the wounds took on new significance. Powerful antiseptics previously had been developed, but they were often harmful or not as effective as previous laboratory studies had indicated. Louis Pasteur had first proven the existence of microorganisms and went on to reveal the germ theory of disease, beginning in the 1860s. This theory held that it is the presence of germs and their replication that causes disease in the body. Joseph Lister further explored ways to stem the spread of germs. He served as professor of surgery in Glasgow, Scotland, and also worked with the local hospital. In the 1860s, Lister took Pasteur's findings a step forward by implementing protocols that used carbolic acid as a cleaning fluid on the wounded. A 5 percent solution of carbolic acid was applied to patients who had healing wounds. The infection rate in the hospital ward dropped dramatically. Lister also required doctors to wear clean gloves for each patient and had medical equipment sterilized with carbolic acid to prevent the passing of infectious organisms to patients. These ideas were controversial among the medical staff, who tended to hold to traditional methods, but the results clearly endorsed Lister's methods. His work gave rise to the advent of sterile surgery. Lister died in 1912, but others continued to refine his methods and search for even lower infection rates.

A research center in Boulogne, run by Sir Almroth Wright, showed that concentrated saline solutions produced less infection than some antiseptics.[25] Initially, at the beginning of the war soldiers were supplied with iodine solution as a disinfectant. It did not prove to be very effective, with one doctor detailing that it "has proved a failure in treating large wounds."[26] There were various concoctions of both liquids and paste that were suggested as the war progressed—mercury perchloride, carbolic acid, hydrogen peroxide. There were those, such as Wright, who advocated that the body should be able to heal itself and were staunchly against antiseptic solutions. Others likewise championed the use of antiseptic treatments as the best way to stem the numbers and rates of infection that were being seen among the wounded men from the war front.[27] Reports of success with salt treatments or antibiotic paste made the rounds. Ultimately, iodine use was limited for larger wounds and replaced with one of the major innovations to affect the treatment of wounds in World War I, the Carrel-Dakin method. The method was named after Alexis Carrel, a surgeon who had already spent years studying how to best care for wounds, and Henry Dakin, a chemist. Carrel saw in the war a chance to find new ways to save the wounded and utilize scientific principles and experimentation to find the best way to do just that. Fortunately for him, his work was financed by the well-heeled patronage of the Rockefeller Institute in New York. He also had support of the French government and conducted his research and experiments in a French hospital near the frontlines of the conflict.

The method utilized a specific solution, often referred to as Dakin's solution, which was a neutral sodium hypochlorite. Treatment began with a thorough cleaning of the wound and stopping the bleeding. Continuous irrigation with Dakin's solution through tubes inserted deep into the wound was next, with periodic samples from the wound checked with a microscope. Once the wound was clear of infection, then it would be closed. This multistep process became known as the Carrel-Dakin method.

Doctors found success with it, and its use spread outside of the French military hospital system. The Carrel-Dakin method was officially adopted by the British in 1917 and was implemented by the US Army Medical Services from the beginning of their time in Europe, starting the same year.[28] Flavine and dichloramine, two new antiseptics, were also created to help clean wounds.[29] This pattern of the spread of medical knowledge is indicative of how the medical personnel worked together across national lines to adopt and learn about better procedures.

Doctors obtained these good results with the Carrel-Dakin method; however, it was a time-consuming and tedious process for the doctors to continuously check the wound and to care for it, as it remained open for so long. There were reports of patients having to be monitored and cleaned with the solution not only for days but also in some cases for months.[30] Soon, Georges Gross, a French army surgeon, began to excise damaged tissue from the wound and dispensed with the Carrel-Dakin solution. Widely published in 1915, this treatment evolved into what is known as debridement.[31]

The term *debridement* had been around for centuries, but what eventually came to pass as debridement in the Great War was new. Debridement called for the removal of any foreign object and any damaged tissue in a wound, and for the surgeon to clear tissue to a particular depth of the wound. At times this could be an extensive amount of tissue, as with very deep wounds. And at first this procedure was not universally adopted. Surgeons feared they would accidentally remove part of the healthy tissue along with the damaged tissue and that this would cause even more trauma to the already debilitated patient. However, by 1916 debridement was regularly done on wounds. The next question became when to stitch up a wound.

The issue of how to treat open wounds was not quickly put to rest. In the years prior to the war, many doctors advocated leaving wounds open so that further treatment and cleaning was feasible, while others believed a wound should be closed as soon as possible to prevent germs from further contaminating the wound. The pages of the *American Medical Association Journal* chronicled the debate as first one doctor would write in favor of early closure, and then another doctor would oppose it.[32]

The problem with closing wounds early was that many times they were closed without being adequately debrided (removing damaged tissue), and infectious bacteria remained in the wound. Also, a factor affecting infection rates early in the war was the soil on the Front, as briefly mentioned earlier. The Western Front stretched over land that in peacetime was used for farming. This well-cultivated, fertilized (usually by manure) soil would get into the wounds, causing gas gangrene. Gas gangrene was a particularly virulent form of gangrene that rotted the flesh, necessitating amputation once it appeared, and often was fatal. It was caused by a particular bacteria *Clostridium perfringens* that was in the manure. Gas gangrene, however, was easily diagnosed, for two reasons. One was a crackling sound that the wound made when pressed on. The other was the odor it emitted, which was intense and pervasive because of the decay of the tissue. Doctors were never able to effectively combat this pathogen. It could develop and spread very quickly. One fear was that gas gangrene could take hold in wounds that were closed early and therefore had anaerobic (lack of oxygen) conditions. There was no treatment to reverse it.

Today, gas gangrene is treated with antibiotics, sometimes with a hyperbaric oxygen chamber, and still is often resolved only through amputation. Instead, for the doctors of the First World War, without some of these aids, time helped. As the years of war passed and the land became more and more turned over by the shelling, the fertilized top soil disappeared, to be replaced with more chalky earth that held less of the contaminant manure.

Another danger and complication from the soil came in the form of tetanus spores. During the American Civil War, the tetanus mortality rate was between 89 and 95 percent. Obviously, this type of disease rate was of great concern. But anti-tetanus toxins available at the beginning of the war were not prescribed for every injury. It was not until 1916 that the anti-tetanus toxin was administered to all soldiers with injuries, whether or not they showed any symptoms. The tetanus rate naturally dropped. By the end of the war, although the mortality rate for tetanus was still high—between 20 and 50 percent—only 0.1 percent of soldiers were infected by tetanus.[33]

Moreover, with time, the wound debate was settled. Debridement was done on all major wounds, although its advent did not diminish the importance of the Carrel-Dakin method. Carrel-Dakin was still used extensively when a surgeon felt that either debridement might not have been complete or to prevent or limit infection when a wound was left open. Wounds might be open because they had not healed or, in some cases, specifically so they could be constantly irrigated to minimize the risk of infection. When wounds were not immediately sutured, this was called delayed primary closure.

An integral part of treatment, mobile laboratories also assisted care of these wounded soldiers. For the first time, bacteriology could be conducted in frontline hospitals. If a wounded person survived surgery, infection was the most common complication. Science had made much progress in understanding bacteriology and disease development. Tests of bacteria to determine how much of it was present in open wounds were made daily to give doctors notice if an infection was beginning. Bacteria counts also had to be sufficiently low before a wound was sutured closed.

Some of the most dangerous (to the patient) wounds were abdominal ones. But abdominal surgery made great advances during the war. Here too the conservative approach dominated early treatment of penetrating wounds to the abdomen. Surgeons felt that these patients would not survive surgery. Instead, the only treatment was to watch and wait. Not surprisingly, mortality rates were extremely high, nearly 90 percent in previous wars. In the first part of World War I, surgeons moved to operate on these wounds and do what good they could for the patients.[34] It was soon found that early operation greatly increased the chances of recovery. In 1918, orders outlined a staggered level of treatment options for these wounds.[35]

Orthopedics

The specialization of orthopedics gained stature in the war. Orthopedic injuries were mostly caused by bullets and shrapnel that broke a patient's limb bones. The early years of World War I saw high mortality and amputation rates for these types of injuries. This was in keeping with what had been seen in previous large-scale wars such as the US Civil War in the 1860s. In 1915, wound mortality approached 28 percent, and the amputation rate was found to be as high as 40 percent for extremity injuries that involved the bone.[36] However, the fact that these rates had decreased dramatically by the end of the war was testament to the advancements of medical science during the war. Medical care focused on preserving and setting the wounded bones though surgery was often required.

Small splints and bandages would be applied by the field ambulance personnel or at the casualty clearing station after a patient was brought in. Once moved to a base hospital, the patient would begin the process of having a more fixed splint or traction put in place.[37] The wound also might result in a surgical consultation.

With the improvements made in surgery and treatment of infection, the overall amputation rate dropped drastically. The consensus became that wounds should never become so infected that they necessitated amputation. Signs quoting Alexis Carrel, who invented the Carrel-Dakin solution, were put up in hospitals. "Every wounded man who develops suppuration [infection necessitating amputation] has the right to ask his surgeon to justify it."[38] The fear of infection was so great initially that surgeons even amputated preemptively on some cases. Coupled with the success against infection was the practice of immediate amputation for all compound fractures, which doctors had also abandoned by 1917.[39]

When a wounded patient did have a limb amputated, the surgical technique was by no means left unchanged through the war. Different techniques developed as to how to best close the wound caused by the amputation. If surgeons had the time, they would employ a flap method. After amputation, surgeons formed flaps from the soft tissue alongside the injured limb. In effect, tissue then extended beyond the amputated bone and could be used to cover the amputation area. This could assist in the healing process and in the appearance of the limb. However, during periods of great activity, this method took too long to perform.

Instead, allied armies began using a quicker, flapless method also known as a guillotine amputation. This method cut the tissue at the same place as the bone. Though faster, the flapless method also required much more post-operative care and corrective surgery. Doctors found when the flapless method was used, the soft tissue further retracted above the bone, rendering the stump conical shaped and unready for any prosthetic device.[40]

The future success of a prosthetic device for those with amputations was the skill and attention of the surgeons. In 1915, a report to the *British Medical Journal* stated that "all stumps are not as good as they might and would be if operating surgeons at the time of the operation could give more thought to the apparatus that will eventually have to be fitted, and during the after-treatment always had the same point in mind, taking pains to prevent contractions and adhesions in the stump and in the joint above, which must limit movement in the future."[41] For many of the wounded, amputation was not necessary, as their injured bones were able to be set properly and healed well, though there was always the possibility of complications. One of the most dangerous wounds to receive was a gunshot wound to the femur, where fatality was common, and severe disability and deformity usually followed the survivors.[42] An article in 1918 advised surgeons that when working with fractures of the leg to be particularly aware of possible damage to the femoral artery, which was often masked by the fracture.[43] Fractures often were treated with the "Thomas splint," which could be used on either arms or legs, in traction or not. The importance of splits was demonstrated by the issuance to British forces alone of more than 1.5 million splints.[44] Doctors worked extensively with different types of splints and traction settings and positions to maximize the body's own ability to heal. The photograph in Illustration 2.1 depicts

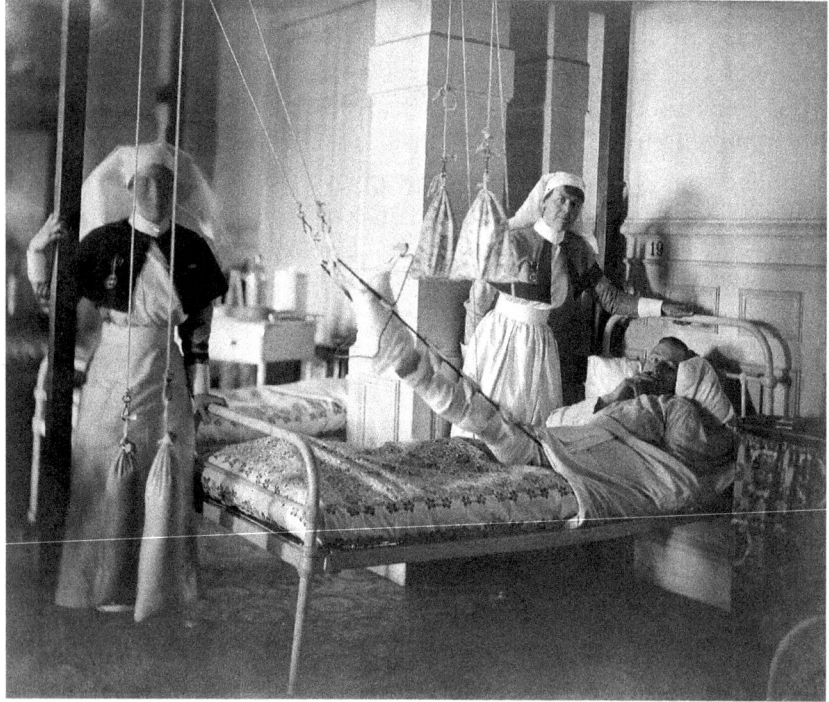

Illustration 2.1 Demonstration of orthopedic splints and the use of pulleys in treating a fracture patient.

Photo Credit: 'World War One, 4th General Hospital, Wimereux'. Credit: Wellcome Collection. CC BY https://wellcomecollection.org/works/e9w8ggtm

a fracture patient in splints and traction, with the tension of the traction coming from the bags with weight in them.

Also important, as many doctors noted, was the continuity of care. In the case of orthopedics, the British focus on emergency treatment had improved the patient's chances when faced with, as mentioned above, a gunshot wound to the femur. At the outset of the war, one estimate stated that 60 percent of such patients died of their wound. By 1917, that number had decreased to 12 percent.[45] However, making sure that patients healed as best as possible meant constantly observing splints and making sure proper treatment was applied. If it wasn't, it could mean later problems with healing or even the inability to be fitted for a prosthetic.

One of the major changes in this specialization had to do with the setting of bones. A doctor with the US Medical Corps observed that "the tendency has led steadily and progressively away from the methods of absolute fixation... towards methods in which the main principle is traction."[46] The descriptions of the problems found early in the war with the use of plaster splints on fresh fractures highlight the troubles in previous wars. The limbs could be constricted causing edema, muscles atrophied, pressure sores developed, it was difficult to keep the wound clean from dirt, and joints stiffened with lack of use.[47]

Instead, by the later part of the war, splints might act more as a cradle for the fractured limb for a hospitalized patient while traction via a pulley-weight system was applied. The goal was to keep the muscles around the fracture at rest (not taut) as much as possible. This meant that not as much force would be needed to keep the fracture set, and it promoted better healing of the injury.

Some traction scenarios were simple, with an arm raised by a pulley. Others were certainly more involved and might encompass the entire bed frame where a patient lay. For example, in a case where both of a patient's legs were broken, there would be sling supports under the torso and each leg, four bars over the top of the bed to handle nine weights, and 12 pulleys to angle the body and limbs, so the bone fragments could heal properly.

Doctors sought to minimize infections with compound fractures and joint injuries when they could, as these types of wounds were often open wounds. They realized that bits of a shell fragment or cloth, if left in a joint or wound, could trigger infections, but they also reminded others that medical equipment used in treatment, if not sterile, could do the same. A drainage tube could prove to be a deadly carrier of disease if precautions were not followed.[48] If the tissue became unhealthy or infected, no attempt at correcting the fracture would be successful.[49]

Transfusions

Major advancements were made in the area of blood transfusions. The premise of blood transfusions had been around for decades. The practicality of them, though, was sadly lacking. Blood congealing when exposed to air was one problem. Another equally serious one was that some people had fatal reactions on receiving another's blood. In 1901, Karl Landsteiner made an important discovery—he was able to differentiate the varying blood types that people had. This made blood transfusions a more viable option. And they were performed in the war as early as 1914, using a particular process. This process, known as Carrel or Crile (depending on which doctor's procedures were followed), suffered from two main obstacles. One, it necessitated a direct transfusion from a donor. Second, it was a difficult procedure that took meticulous care and was very time-consuming. It usually required several medical personnel to work simultaneously to complete the transfusion, which was done via surgery on both the donor and the patient to expose the blood vessels that were to be used in the transfusion. These issues precluded its use in emergency situations.[50]

In the years just preceding the war and in the first years of the war, an indirect transfusion method was utilized. Blood from a donor would be put in a small tube called a Kimpton-Brown tube and was taken over to a patient. The blood flowed out of the tube due to pressurization differences. Later, syringes were also used to collect and then transfuse blood.[51] This improved transfusion logistics some, but it did not solve all the problems transfusions presented.

A critical obstacle to widespread blood transfusions was storage of the blood prior to transfusion. Doctors found that the blood quickly coagulated (formed clots), which prevented its use. Tests done in the US Army in

1916 showed that if sodium citrate, glucose, and salt were added, the blood could be refrigerated for two weeks without coagulation or deterioration. In 1918, Captain Oswald H. Robertson formed the first storage systems for refrigerated blood. This was the start of blood banks. The transfusing of blood never became fully routine during the conflict, but it did greatly affect the way surgeons dealt with patients. The United States and Great Britain established transfusion resuscitation teams that were attached to special shock centers and often deployed in field hospitals.[52] By the end of the war, some doctors were even using as blood donors men who had been gassed but who were not in the acute stages of gas poisoning.[53] It was only in retrospect after the war that the medical community began questioning blood transfusion as a method of spreading disease. The first cases of donor blood spreading disease were documented during the years of the war in civilian medicine, but by the last year of the war, doctors on the front only chose to actively avoid donors who had syphilis and malaria.[54]

Aside from the availability of fresh donor blood, another obstacle to performing transfusions was identifying the type of blood the patient needed. Wartime surgeons showed reluctance in taking the time (which might be from several hours to a day) to complete blood-typing tests. Doctors felt the patients needed transfusions immediately due to their conditions, and they risked the often fatal effects of blood incompatibility. Encouraged by the wartime environment and the pressing necessity, doctors in hospitals in the United States and France developed almost identical, new, rapid blood-typing tests that could be completed in a matter of minutes.[55] This made blood transfusions a true option in medical care.

Still, giving the pressing concerns of medical care in war, often blood tests were performed to decide which donors were eligible rather than trying to match patient to donor explicitly. Doctors had already discovered before the war the unique properties of type O blood, namely that it could serve as universal donor blood to people of all blood types. So, doctors sought out those with type O blood. Often these might be other patients who had not been badly injured themselves. Other times, incentives were given for donors, such as the British Army extending extra leave time for those O-types who donated.[56]

Transfusions were especially helpful when patients were in a state of shock, a common ailment of the wounded in this war. With patients in some cases waiting days to be carried out of frontline conditions, it was no surprise that many were in shock by the time they arrived at a doctor's care. Casualty clearing stations reported that the main reason for shock was severe loss of blood. That state of shock rendered the patient inoperable, and in fact, treatment of shock was usually the first thing done for patients. Transfusions allowed the patient to become stabilized so that necessary operations could proceed.[57] As transfusions became more routine and the ability to store blood became more understood, blood became available even in frontline aid posts that normally did not have the most up-to-date supplies and care.[58] However, doctors did self-impose limits on how much blood would be transfused. At a maximum, 1,000 ml of blood would be given to a patient. This corresponds to about two pints of blood, when there are about 8–12 pints of blood normally in a human body. In modern parlance, this amount is just over two units of blood. So, in emergency procedures today, a patient may well be given much more blood via transfusion than was done during the Great War.

The use of a blood transfusion depended further on whether a surgeon had the knowledge and expertise to perform it, as there were no set guidelines. In early 1918, an interallied surgical conference was held that devoted itself to the subject of blood transfusions. Surgeons from the United States, France, Britain, Belgium, Italy, Japan, Portugal, and Serbia attended. The conference showed that implementation of blood transfusions was becoming more commonplace.[59] During a high activity period in 1918, an American surgeon reported "a casualty clearing station might see 35–50 transfusions in a 24-hour period."[60] Again, the interconnectedness of the medical community is seen in conferences such as this. As with many procedures, transfusion care became more streamlined in the later stages of the war.

X-Rays

Assisting doctors and nurses in determining the proper course of treatment was the use of fairly new technology—radiology, or as it is more commonly known, X-rays. The discovery of this form of radiation by Karl Wilhelm Röntgen in 1895 would deliver a great service to medical practices. Röntgen observed that these rays passed through some human tissue, but bones and pieces of metal were visible with the rays. Although something of a sideshow stunt for many, X-rays quickly proved their worth to the medical community as a diagnostic tool. One of the first military uses occurred in the late 1890s as doctors used X-rays to locate a bullet in a wounded soldier.[61] But the logistics of using the technology proved to be a problem. In the Spanish-American War, an X-ray machine was used but took over half an hour to warm up.[62] Using X-ray equipment was also a delicate proposition—it required a high electrical voltage, which could break down some of the X-ray tubes used. Shortly before the war, a new product, a high-vacuum tube, that could handle higher voltages became available.

X-rays served many functions in the Great War. They allowed doctors to pinpoint where a projectile was in a body. If carefully analyzed, the X-rays would show not only location but also the depth of an item in tissue prior to surgery. This would allow for faster and more localized surgeries. Because there was always the possibility of a great deal of shrapnel in the wound given the weapons of the period, identifying all objects that needed to be removed was of the upmost importance. Healing would be faster for the patient, and the likelihood of infection would be decreased. X-rays would also be used regularly to determine the type and extent of injury, particularly with broken bones. Bone fractures were a common occurrence with penetrating shrapnel wounds, and X-rays definitely helped surgeons as they prepared for what they might encounter in a surgery. It was only in times of great activity, and then only in cases where surgeons were certain of a bullet's location, that X-rays were rarely employed.

René Van Tiggelen explored the issue of radiology use during the war in his book *Radiology in a Trench Coat*. He noted that the French had pioneered mobile radiology units before the war. However, as happened many times, the full scope of the need for this technique was not comprehended as the war began. "Many of the temporary hospitals, both military and Red Cross, have no electricity," which precluded the use of X-rays.[63] What eventually was developed across the various medical services were ambulances with electric generators to power the X-ray machines. These generators were even powerful enough to produce sufficient electricity for a mobile hospital unit.[64] In addition to mobile units, bedside X-ray machines were also developed. There were at least 528 X-ray outfits among the Allies during the war, with 14 of them being mobile units.[65] The cross-pollination of knowledge across these X-ray units was immense and continued through the duration of the war.

Once the problem of setting up an X-ray machine was solved, another presented itself. Due to the newness of the technology, there were few people trained in how to administer and read X-rays. This problem existed among the European Allied medical services and the US Army Medical Service once the United States was in the war. In the United States, the American Roentgen Ray Society played a role in recruiting doctors into training programs. At Fort Oglethorpe, Georgia, at Camp Greenleaf, physicians went through a two- or three-month program to learn the techniques of radiology. Van Tiggelen mentions that over 700 physicians were trained in these courses, and the US Army was insistent that X-rays be under the control of physicians, not technicians, which meant that after the war there was a large cadre of professionally trained radiologists, and the fact that they were physicians increased the standing of this medical specialization.[66]

There were also additional programs to train technicians. These were mostly in major metropolitan areas; as Van Tiggelen notes, they "opened in Baltimore, Boston, Chicago, Chickamauga (GA), Kansas City, Los Angles, New York, Philadelphia and Pittsburgh."[67] Medical personnel were trained both in radiology and fluoroscopy. Radiology involved the taking of X-ray sheets that could be later examined. Fluoroscopy involved the examination of a patient via an X-ray machine, but it does not involve taking actual pictures. Fluoroscopy would be used as a diagnostic tool during a patient's evaluation.

The US Army Medical Service published a manual to assist in the training, but it also served as a resource during deployment. The *United States Army X-Ray Manual* described the science behind the X-ray machines and how the machines were to be set up and operated. It explained the various types of X-rays that should be taken for the myriad of injuries that might be seen, discussing in detail certain conditions that X-ray operators should be aware of and even describing conditions that can cause an X-ray failure. This is when the rays fail to penetrate to render an image. The manual was incredibly comprehensive as it dealt with not only traumatic injuries but also chronic diseases that might necessitate a radiological consult.

The manual contained information on the safety issues surrounding radiology both for the patient and for the operator. The manual noted that there were some things an operator can do for self-protection, but given the realities of war, the only one that could truly be implemented was the practice of having a layer of protective material, normally lead, in between the X-ray tube and the operator. It also counseled operators to regularly check their equipment for any leaks or loose joints that could increase the level of exposure.[68]

By the time of the war, the dangers associated with X-rays were well documented. Many of the early radiological technicians developed cancers due to their prolonged exposure to the invisible rays. This was particularly true for technicians' hands as they sometimes had to hold a patient in place during the X-ray itself.

For the patients, exposure to the rays would obviously happen without any protection, so managing the intensity of the rays and the exposure time was important. Generally, depending on the intensity, anywhere from 12 to 20 minutes of exposure could cause what was termed erythema. This was a redness of the skin after X-ray exposure. The skin had literally been burned by the exposure, and the patient was not to have X-rays on that area for three weeks.

What is now known regarding X-ray exposure for patients would reduce the amount of exposure time advised in World War I, as people now would rarely be given the amount of radiation that was fairly commonplace during the war.

One area that received more attention at the radiologist's hands was the skull. The manual advised, "On account of the many difficulties that are encountered in the correct diagnosis of head lesions the routine X-ray examination of the skull is of the greatest importance in all head injuries."[69] Patients with head wounds were often not very conscious, or if they were conscious, they would at the least be "irritable."[70] Radiologists were cautioned to expect this and to be extremely careful to move the patient as little as possible. The problem with X-rays of the skull, though, was bone density. The denser the material, the more current had to be used, or a greater amount of time for exposure had to be factored in. Keeping a patient still for even longer periods of time could prove quite difficult.

Different countries had different medical approaches to X-rays, though all used them. For example, the United States routinely put patients arriving at base hospitals through X-rays in their initial assessment. The British and French Medical Department instead conducted clinical assessments first, and only if indicated had the patients proceed to the X-ray department.[71] The French had not seen the military medical necessity for X-rays initially in the same way as the British, but they soon did. In the early years of the war, Marie Curie, the famous French scientist and Nobel Prize winner, personally donated money for X-ray equipment and encouraged others to do so. Both mobile and stationary units were equipped by her efforts.[72] Curie saw that a mobile X-ray unit deployed into forward positions would be of great help before surgeries commenced, and started her personal effort to provide X-ray units in the fall of 1914. Each mobile unit she designed contained its own generator to power the equipment, and these units were soon known as "Les Petites Curie." Curie herself, working with the Red Cross, drove units to field hospitals, helped set them up, and began a training program for other women who would serve as technicians. Eventually, Curie created 20 mobile units, and she oversaw the establishment of 200 radiological stations at field hospitals.[73]

Plastic Surgery

Ultimately, all of these medical advancements meant that patients were surviving injuries that in prior wars would have killed them. This also meant, however, that patients were now having to cope with what, in many cases, were debilitating injuries. This was especially true of head wounds. These severe injuries and patient survival of them gave rise to a new field of surgery—plastic surgery. Properly termed *maxillofacial surgery*, plastic surgery became a necessity for many of the survivors. Of all the achievements of modern medicine during the war, "none is of more vital importance than the process of restoring faces."[74] More times than not, injuries were major. Facial bones had been completely shattered. Eyes or ears were missing. Jaws were completely blown away. These soldiers were lucky to be alive, but to regain a semblance of a normal life, the plastic surgeons were essential.

At the beginning of the war, such advancements in this specialization seemed improbable. There was simply so much that surgeons did not know. In his contribution to the book *War Surgery 1914–18*, John D. Holmes notes that much of the understanding of blood vessels and bloody supply at the detailed level necessary for successful plastic surgery procedures had only just begun to be explored in the late 1800s.[75] In *A History of Military Medicine*, Richard Gabriel and Karen Metz point out that surgeons were so inexperienced in treating maxillofacial injuries that "patients suffering from injuries to the face were often transported in the supine position, a position that blocked their airways and produced death."[76] However, Great Britain soon started the first maxillofacial hospitals, and when the United States entered the conflict, it quickly established such hospitals as well.[77]

Plastic surgery was able to effectively draw from the few principles that had been established in it before the war, but it did innovate some new techniques, including skin grafts, treatment of burns, and the new use of skin flaps during reconstruction.[78] The plastic surgeons benefited immeasurably from the amount of practice they received during the war and were able to better assess the feasibility of certain procedures. Surgeons were able to learn exactly which techniques afforded the most benefit. Early intervention was found to be key for a good prognosis. Obviously, this could not always occur, but moving forward in the years after the war, the need for this type of intervention could be better argued for and implemented.

As a general rule, plastic surgeons did not debride facial wounds because gas gangrene in the face was an extremely rare occurrence. Instead, all tissues that could be preserved were. Doctors did strive to have the wounds drain properly. They used facial supports to relieve muscle tension and support the tissue until sutures could be removed. Splints were also used to hold fractured jaws in either open or closed positions to facilitate healing.[79] In another departure from regular surgical treatments, doctors also found that use of the Carrel-Dakin solution was not a good choice for facial injuries, and they avoided using it.[80]

Tied in with the new design of surgical procedures was finding the best antiseptic solution to be used around and in wounds of the mouth. Reporting from a hospital ship, Dr. A. R. Fisher noted that he had been using chloramine as an irrigant for these types of wounds and found it to work better than other antiseptics. It was less harsh on the tissue, for example.[81] Various solutions were used throughout the war.

One of the key elements of plastic surgery was to make sure function was restored first. Often plastic surgeons were not available in forward positions. In these cases, their colleagues, the dental surgeons, stepped in to administer what help they could. Even in positions away from the front, the work of dentists was integral to the repair of jaw injuries.

Modeling casts were used to build dental appliances for work done in the jaw area. These appliances might be permanently affixed in the mouth to enable a more normal shape and contour of the face. Others were worn only for weeks or months to slowly stretch the muscles that controlled the jaws. "The average case showed from 2 to 4 mm. improvement per day" due to such stretching.[82] At other times surgery was necessary to remove the source of the restriction for movement.

Treatment often took place over a protracted time period. Multiple surgeries were needed to accomplish both infrastructure repair of the jawbones and facial bones. Further surgeries then tackled the cosmetic appearance of the patient. Sometimes this was accomplished by medical staff in England, for example, after a patient was evacuated home. Other times base hospitals on the Western Front conducted the care.

Once the bone infrastructure was ensured, the medical staff would move on to soft tissue repair. The use of grafts of free skin to replace mucous membranes of the mouth usually either failed before the war or was itself considered only experimental. During the war, however, it became established "as a definite procedure for which an almost positive assurance of success could be given."[83]

Again, modeling compounds and casts came in handy. Tissue to be used in skin grafts was placed on modeling casts made of each patient's face in the afflicted area. The skin could then be stretched and shaped to better cover the area to which it would be applied on the individual patient.

This procedure involved then attaching the skin graft while leaving the compound exposed to keep pressure on the wound. Depending on where the graft was needed, this could be very uncomfortable for the patient. American maxillofacial surgeons usually left it in place for 10 days and then replaced the compound with a piece of rubber to keep up the pressure and stretch of the graft.[84]

New innovations in one specialization also were applied, when they could be, to other ones. Doctors found that the soft tissue, for example, around the mouth often thickened and became less elastic after a wound occurred. They experimented and discovered that intensive X-ray treatments often softened the tissue to allow better surgical results. Patients might only have to undergo one treatment, but they might be given as many as three.[85]

For many of the soldiers, follow-up surgeries were inevitable. Reconstruction hospitals on the home front continued treatment. Some patients were not able to even begin treatment until nearly a year after the injury occurred. For US troops, treatment at Walter Reed in Washington, DC, was common. Healing took months to accomplish, especially in cases where bone grafts were used. Bones from the pelvis and from the ribs often provided the material used for the rebuilding of jawbones. Such destruction of the face was often caused by fragments of high explosive shells. After grafts were inserted, splints of varying types were used to hold the area stable for the bones to grow together. In general, splints were used for three to six months following the surgery.

Other types of jaw injuries required more simple prescriptions for recovery. Tissue damage and scarring could affect the muscles and nerves of the jaw area, prohibiting patients from being able to eat and chew their food. Devices as simple as a clothespin with a spring were used to stretch the muscles in such cases.[86] Devices with screws were also used, not unlike how doctors today use medical screws in the lengthening of leg or arm bones.

Various types of headgear were also utilized to provide support and stability for the procedures during the healing process. Lines for traction were also used for patients. Although this is a treatment that is more related to the healing of arm and leg wounds, it was also used in relation to head and neck injuries. Simply getting a patient to a point where the jaws or nasal passages were useful again was not enough. Further work was often done to minimize deep scarring of the face and improve appearance.

Even what might look to the uninitiated as a straightforward repair could become quite complicated because of the anatomy and musculature of the face. Patients who were believed to have a foreign body in their facial tissue were X-rayed so that the medical staff could determine the proximate location of the bullet or grenade or high explosive fragment. Ideally, this would be done as soon as possible after an injury. This became more complicated if the wound healed prior to the patient being examined. Given the location of the maxillofacial units, that was a real possibility.

X-ray units were only mobile units at the beginning of the war, but as their use expanded, radiology became available in many of the treatment centers for the wounded. Specially trained radiographers followed early principles of the process to aid in the investigation of the wound. Most often doctors used a fluoroscope to better pinpoint the foreign bodies. This was especially helpful when numerous shell fragments might be present. The fluoroscope utilized the principles of X-rays in a real-time evaluation of the body, providing better depth and detail to allow for more precise excising. This was key when plastic surgeons were striving to retain and protect as much tissue as possible.

If a wound had healed and the surgeons wished to improve the appearance of it, they would begin by excising the area and removing the underlying scar tissue. If the area had a depression due to previous loss of tissue, subcutaneous fat or abdominal fat would be used to fill in the area.

Skin flaps were used to close wounds and also to hold in place displaced tissue such as a drooped eye or a lopsided mouth. For wounds on the lower face, skin flaps, called pedicles, were prepared. These were taken from the neck and were reoriented while still attached and turned upward to cover the area up from the lower jaw. As the skin grew over the wound, it was separated from the rest of the flap and the remaining skin flap could be returned to its original location even. This could actually be done with large segments of skin, even coming from the chest, which made it a very versatile procedure.

Another not uncommon reconstruction involved the nose. Cartilage from a rib was used and often transplanted to the forehead, where a flap of skin would be used to rebuild the nose. A flap from the temple was often preferred, but depending on the injuries, it was not always possible to do this. Cartilage grafts could take a long time to heal and were considered a serious surgery. The US Army ended up designating specific hospitals for such patients. Selected maxillofacial and dental surgeons would be assigned there. Medical staff found that also having specially trained dental officers accompanying them on lengthy transports of patients helped as well. The British similarly had specialty facilities for this growing area of expertise.

Despite the best efforts of the surgeons and medical staff, some injuries defied repair. In those cases, another option was possible. Partial masks were made to hide the injury to the face. These masks were particularly geared to injuries that had affected the eyes and cheekbones. Often the mask was attached to a fake set of glasses, so the mask could be held on the face by the earpieces. Other designs did not incorporate the earpieces. Anna Coleman Ladd was particularly gifted as an artist and began in 1917 to sculpt actual masks for those whose injuries surpassed even what these medical innovations could help with. She not only sculpted the features but painted the masks to blend in as near as possible with the patient's own skin tone and features.

It should also be noted that some of the procedures that were found successful in treating facial wounds would also serve to enhance treatment of other types of wounds. In particular, the work with skin flaps became important for orthopedists as a key part of healing after amputation. A well-healed area was important for a good-fitting prosthetic device for a wounded patient who perhaps lost his arm or leg.

Still, the resiliency of the human body to repair itself was amazing. In his book *Stretchers: The Story of a Hospital Unit on the Western Front*, Frederick A. Pottle remembered "one remarkable case of a man who walked into the operating room without help. A bullet had gone clear through the front part of his head from one cheek bone to the other, below the eyes, and above the teeth. There proved to be nothing for us to do at all. The wound was clean and healing of itself."[87]

The doctors themselves realized the specialized nature of the work. Writing in 1916, one dental surgeon noted regarding gunshot wounds to the head and face that they occurred in "no means a negligible quantity, and their efficient treatment is somewhat out of the sphere of the general surgeon."[88] In the years after the war, plastic surgery refined its techniques and emerged through the twentieth century as a specialization that was needed by civilian society as well.

Although plastic surgery is often thought of in terms of improving appearance, there is also the very practical side of the specialization—to improve the function of the face and jaw. The medical staff that worked on these patients not only afforded them a more functional existence, but they made it possible for them to reenter society with less anxiety over their appearance. Patients regularly expressed great appreciation for the work done.

Conclusion

The Great War's tragic carnage necessitated a standardized response from the military medical departments of the Allied countries involved. The medical staffs of the Allied countries were able to make significant gains and advancements in their treatment procedures that gave patients a better chance at life and recovery. The reality is that this knowledge came about because of the extensive number of patients seen and the ability to discern through practice what did and did not work best. By the later stages of the war, much of this new knowledge became more regimented and dispersed through manuals such as those put out by the Red Cross to instruct on exactly how certain injuries should be tended. Regular reports from doctors also appeared in journals on both sides of the Atlantic, such as the *British Medical Journal* and the *American Medical Association Journal*, as they sought to inform others of what discoveries they were making and what treatments seemed promising. The stationary nature of the Western Front also contributed to the efficiency of the medical care and the ability to have better outcomes with early treatment, as fixed hospitals were possible in these circumstances. The medical bureaucracy that facilitated the growth of the medical specializations in the war could be seen in other areas, not the least of which was the fixed use of patient charts.[89] Doctors could mark what the patient was suffering from, what they had done in treatment, and what medicines had been administered. This information would then be easily accessible to consult for any subsequent medical visits.

Though doctors learned new procedures and made new discoveries during the war, they also learned the valuable lesson of when they should not intervene. For example, emphysema (fluid in the lungs, as it was described then) was a frequent complication of influenza. It was usually treated by surgery, which resulted in a mortality rate between 60 and 85 percent. The surgeons learned to stop performing the surgery, and the death rate fell to only 9 percent.[90]

As Perrin Selcer observed in his account of the development of the Carrel-Dakin method, there was a public relations side to all of the medical developments and innovations in the war. Although these soldiers might not return quickly to action, it was important for public morale to know that everything that could be done to heal them was being done. Seeing that "cures" were possible for the wounded was key in continuing to have support for the war on the home front, and it was an encouragement for the soldiers still in the line.

3

Medical Personnel

Medical staff on the Western Front served as a key ingredient in keeping the soldiers fighting. The organization of the medical care that was offered came from both military medical departments and volunteer outfits that mobilized as soon as the war started. The French and the British obviously had the most established presences as the war progressed, but a steady stream of volunteers also came from across the Atlantic, even in the early months before more formal structures were in place. Many categories of positions existed within the medical community and, as the previous chapter discussed, increasing specializations. Further cadres of nurses, stretcher-bearers, ambulance drivers, and the like all performed work to assist patients. Without all of their efforts, the armies could not have stayed intact and continued to fight.

Memoirs and accounts of those who served in medical units often began in the same tone as any memoir later offered by a soldier. There was a sense of excitement, patriotism, anticipation, a feeling of serving one's country and one's fellow countrymen evident in the prose. There was a desire to contribute. And there was often an awareness of adventure being ahead of them. But actually serving on the Western Front would be a bewildering experience for many in the medical units, just as it was for the millions of soldiers. The number of wounded could be overwhelming; the logistics of getting them to medical professionals could be overwhelming; and the fact that all people must continue with their jobs, even as their own lives were at risk, could, too, swamp one's senses.

A few accounts, mostly written by doctors, appeared during the war. Others appeared after the war and contained more realistic expressions as the authors processed what the war had done. There were, though, a few critical accounts that appeared while the war was still happening that highlighted the human tragedy that was unfolding in the hospitals and casualty clearing stations.

Medical Corps—General Experiences

The medical corps of a military encompassed all the personnel who were in military service and attached to the medical units. For medical officers the organization and training of the unit that would proceed to the Western Front consumed numerous hours. For the British Expeditionary Force, medical officers came not only from Britain but from all parts of the British Empire, as did the troops themselves. In the Canadian Army Medical Corps, officers were responsible for making sure that new recruits to the medical unit learned how to drill properly and were trained in medical work, such as first aid, all before departing across the Atlantic. These new recruits might come in with absolutely no medical training, so their entire education had to be built from scratch.[1]

People chose to go into a medical unit without any previous training for a variety of reasons. Some may have been contemplating a medical career of some type, and this would provide a good training ground for that. Others went in because a friend had gone into such a unit, and it seemed reasonable to follow that lead. Some felt a calling for the humanitarian cause of treating the wounded, thinking that this is where they could do the most good. A few undoubtedly went in thinking that being in the medical corps might be an avenue to avoid frontline combat. But these people were certainly few and far between, as the numerous sources attest. The vast majority of those in the medical corps took their work seriously and approached their assignments conscientiously.

Organizing the units involved weeks of training at a home base. Once the units were organized, simply getting to the war could be complicated.

For British physicians, nurses, and the like, it was a short voyage across the English Channel to reach France and begin the journey to the hospitals and the front. This journey was usually by train and could take some diversions depending on the war and military needs. Trains could be rerouted. They could be backed up. And with such increased use, they could break down. A trip that may have taken less than a day before the war might take several days after the war started.

As will be explored shortly, Great Britain was unsure how many doctors and nurses would be needed, and did not want to overcommit to a large number that might not be needed in the end—the expectation being that this would be a short war, after all. In stark contrast stood the United States, which had the benefit of hindsight by the time it entered the war in 1917. The United States knew it was underprepared and therefore sought to have even more medical personnel ready for service than were thought to be needed.

The organization of base hospitals in the United States was initiated sometimes by local doctors and other times by the military. Generally, the assigned staff of a base hospital would number over 200. The history of one base hospital detailed how people were examined as they became assigned as staff to the hospital. They had two weeks of quarantine to make sure they were not suffering illness. They then had screenings to assess heart and lung function, feet, other organs, and "a special examination with regard to mentality." They had their throats cultured, and results had to be clear for them to receive a certificate for overseas fitness of duty.[2]

For those coming from the United States, it was obviously a more involved trip, taking approximately a week for ships filled with military personnel and volunteers to make it across the Atlantic Ocean. Dr. Clarence Benjamin Francisco kept a diary during his crossing in May of 1917, shortly after the United States formally entered the war. The ship he was on left the East Coast of the United States on May 19 and didn't arrive and drop anchor in England until May 27. The journey started out rather upbeat, with people getting to know one another and most of the ship being military (rather than civilian) passengers. As the days passed, those onboard would see military destroyers at sea and their own ship do target practice with its guns, and they would have daily lifeboat drills. The rumor of a periscope and its attached submarine would cast a pall over the passengers. Seasickness visited some of those onboard, and the cold weather on a North Atlantic crossing did its part to keep morale down. Some nights, people slept dressed with life belts nearby as a precaution against the threat of the unwanted torpedo. Others were unable to rest at all and stayed up through the night arriving in England bleary-eyed and sleepless. Dr. Francisco described how tense people were onboard as they entered the war zone and how relieved they were when it seemed the danger had passed and the ship was near enough to land and anchoring that people could breathe easier.[3]

The logistics of dealing with the beginning of a war, the number of people involved, and the bureaucracy that needed to be navigated could prove frustrating in the best of times. Early in the war, one doctor serving with the British related how he traveled across the English Channel on a ship filled with the personnel for a new and badly needed base hospital in France. However, none of the equipment for this new hospital had been sent, much to the consternation of the commanding medical officer. "In spite of his utmost endeavours he could not get his hospital equipment on the *Cestrian*. All of the instruments, dressings, and X-ray apparatus had been left behind for another boat, and he thought that he might not be able to get them for another week, or perhaps longer. This was but another example of the lack of control…" As it turns out, the ship in question sailed to France with empty space in its hold.[4] The red tape of bureaucracy was a regular burden the medical personnel had to become accustomed to.

As units reached Great Britain from North America, they went through training with their entire divisions, including war games as practice for what was to come. Their medical officers accompanied them through this training period, treating the odd injury or illness that arose among the soldiers. Medical officers had to make daily camp inspections as well, either at a rear location or near the front. Sick parade was a regular feature of both camp life and combat life. It was usually held early in the day as doctors would assess the current batch of patients and decide what the next stage of treatment would be, or even if evacuation of the patient was needed.

Doctors also found themselves dealing with logistics if they had time to, even when they were assigned to a position near the front. Captain Harold McGill related how he was concerned that transports of wounded were having trouble reaching field posts. He did not receive any satisfactory answer to what was happening. It became necessary to walk the terrain himself to see what impediments might be delaying the carriages and therefore the treatment of the wounded.

The accounts by doctors early in the war often saw them very near the front and in the midst of the fighting. An account left by a British regimental surgeon, Robert V. Dolbey, describes tending to patients and running across an area where a sniper had been firing on troops. Telling his stretcher-bearers to take cover, he hid in a ditch near some rushes at a creek, then ran to protection by a bridge and then to a cottage, to take care of the wounded there before repeating the harrowing trek to rejoin his men.[5] He spent the rest of the day under heavy fire, treating over 150 wounded men as "rifle bullets cracked like whips above us."[6]

Dolbey later described being in a hay field that was still being strafed with fire and trying to tend to the wounded. "Times without number machine guns opened fire, stretchers had to be dropped and cover taken by lying flat."[7] Although not all doctors might be in the midst of active firing to the degree Dolbey described, many found themselves inadvertently in such a situation simply because of the lines moving (particularly later in the war) coupled with the close proximity of the opposing sides to one another.

Those assigned to details such as field ambulances had to move with the army units. This meant long marches sometimes, for the medical personnel just as for the soldiers. One doctor recalled in his account of the early years of the war, "When one is 'soft' and not accustomed to long walking, a day's march like this proves a torture… Towards the end of the long, long day, and in the darkness of the night, with feet swollen and sore, brain and body numbed with fatigue, one did not march, but only stumbled and lurched along the never-ending road like a drunken man."[8] Such proximity to the front could cause the medical corpsmen to encounter Germans, as did a Royal Army Medical Corps (RAMC) corporal who inadvertently stumbled on nearly a dozen German soldiers. Instead of them attacking him, as he believed would happen, they surrendered to him. Considering he was unarmed, as he was in the RAMC, he was quite proud of the turn of events.[9]

The threat artillery posed to the armies was another matter. With effective ranges measured in miles rather than yards, German artillery could inflict massive damage even at positions more to the rear. Often villages in either France or Belgium suffered destruction while being used as medical centers. Medical personnel found themselves treating civilians as well as soldiers in these instances. It was not uncommon for farming families to arrive at a medical facility for help, particularly if there was an injury to one of their children. The Canadian officer Captain McGill detailed some of these civilian cases. In one of his circumstances, the family would alert the unit, and McGill would go out to the family to assess the patient. Again, the cases involving children are the most poignant. There was a seven-year-old he was called out to treat in his first months in France. Then a Belgian girl who sustained an injury to her scalp from a shell fragment was also another civilian victim needing aid.[10]

Still, the examples of civilians turning to military medicine for aid belied the often hostile reaction civilians had toward the military encroachments of their farms and villages. Aside from the obvious disruption to both that the war caused, the civilians also had to deal with the attitudes and behavior of the soldiers, which were not always positive. Soldiers might loot items from a nearby residence, or the military might commandeer property for its own use. Neither example encouraged local civilians to look at the military benignly.

One medical officer described venturing into a farm region near the front and encountering an aggressive dog at a family's home. He described the farm family as follows: "There is no doubt that these people were evilly disposed towards us; yet here they were allowed to remain a few yards from a road within a rifle shot of the German line, and along which a brigade had marched in close column of route a few nights before. No wonder that we were regaled daily with the most plausible and circumstantial stores of espionage."[11] These civilians saw an unwanted war descend on their place on the Earth, and their towns became filled with people from foreign countries. Old allegiances surely died hard under such circumstances.

Many of the doctors on the front kept detailed diaries and journals describing some of the cases that they saw and how their days progressed. Some accounts are very detached and focus almost exclusively on the practice of medicine. Others are more reflective and give a more visceral and emotional description of the author's time in war. As is true even in current times, it was the more exotic injuries that tended to get recalled in the memoirs left by the medical personnel. One doctor relayed cases he saw for the first few weeks he was in France, noting how many eye injuries there were from shell fragments and bullets, and recalling one patient who had, from the force of an explosion, mud driven under his eyelids that packed all around his eyeballs. Amazingly, this last patient made a full recovery.[12]

Harold McGill left one of the most thorough accounts by a medical doctor on the front, and he, too, related cases that seemed to defy nature—as so much did in this war:

> One of the walking wounded was a French Canadian… when he came in out of the darkness he presented a rectangular piece of shell casing of perhaps a couple of cubic inches embedded in his forehead, and projecting like a steel horn… I took hold of the fragment with a pair of heavy bull dog forceps and attempted to extract it. I might as well have tried to remove a railway spike from a sleeper, and had to desist. I asked the boy, "Am I hurting you much?" "Oh," he replied, "I 'ave one head-ache." The soldier walked out with the steel still in his forehead.[13]

For all the advancements, innovations and specialized training that doctors had and displayed during the war, there was also the reality that in certain circumstances medical staff were few and far between and overwhelmed. In Ellen N. LaMotte's *The Backwash of War*, the author, a nurse on the Western Front, gives a brutally honest portrayal of conditions and patients in the hospitals at which she worked. In a poetic style she discusses the issue of personnel shortages:

> Young men, just graduated from medical schools, or old men, graduated long ago from medical schools, were sent to learn how to take care of the wounded. After they had received a two months' experience in this sort of war surgery, they were to be placed in other hospitals, where they could do the work themselves. So all these young men who did not know much, and all those old men who had never known much, and had forgotten most of that, were up here at this field hospital, learning. This had to be done, because there were not enough good doctors to go round, so in order to care for the wounded at all, it was necessary to furbish up the immature and senile.[14]

One surgeon lamented how some patients died too quickly, and others took too long to die. But he felt fortunate that he never ran out of morphia (morphine), and chloroform never seemed to run out, as that was all that could calm some cases.[15] "One felt that one was very glad to be so close up and to be so helpful, and yet one felt so strangely helpless. There was so much to be done, and so many for whom surgery could do so little…"[16]

The vast majority of doctors who served on the Allied Western Front were male. However, there were some female doctors who went to France, though not with the official military medical corps. These female doctors went as volunteers.

Volunteering

As soon as the war broke out, nurses, doctors, and even those without any medical training began volunteering for medical work in the war. Positive propaganda encouraging general enlistment spread to those pondering the medical corps. Volunteering and serving, though, were not as simple as one might think, even in the face of this calamitous war. In Great Britain, the general expectation when the war began was that it would be of short duration, so just as no one talked about a general draft being needed in 1914, so, too, did no one talk about taking all into the medical corps who wanted to be there. The British military was stingy with whom it allowed in the medical corps in 1914. Physicians might be turned away for being too old, for example. The military wanted young doctors who would serve wherever they were sent, but even then, many were put on waiting lists and not immediately incorporated into units. The military was not sure just how many medical officers it would need, and didn't want to oversupply the role. Doctors were told instead that serving the civilian population on the home front was just as necessary as serving in the war on the front lines.

Some doctors did take on extra shifts and responsibilities at home to make up for the young doctors the military did accept. Still others chose to volunteer with organizations such as the Red Cross, which did take older doctors that the military had turned down. Early reports, though, from the fighting in France in 1914 clearly showed there was a shortage of medical personnel to treat the great number of wounded that started to pour into field hospitals. Yet, the War Office in Britain held back in soliciting more doctors. Army volunteers were still easily found for the British Expeditionary Force, and it may have been thought that the numbers needed for the medical corps would similarly simply appear. However, in 1915 the War Office saw the futility of this and began coordinating a more organized approach to obtain the doctors they needed.[17] As the war progressed, such rules on age relaxed, and by 1916 even older physicians were readily welcomed. Once this happened, doctors could transfer from work with the Red Cross, for example, directly into the RAMC.

The efforts at recruiting volunteers proved problematic through much of 1915. There was miscommunication, slow communication, and confusion on the part of those doctors willing to serve but thinking they were not needed, and with the War Office determining just how many doctors they did need. At that time the War Office wanted those under the age of 40 to serve abroad and those over the age of 40 to serve in military hospitals in Britain. Yet, a doctor might apply to serve abroad and never hear back that he was needed. So, he would take on a civilian practice or appointment. Then he would be conflicted when contacted by the military with a commission (perhaps months later), and might not feel he could abandon his patients and current commitment. Many civilian hospitals in Britain faced their own doctor shortages as the war continued and more physicians went overseas.[18] There was also much debate about how to handle doctors leaving a private practice because most likely they had to buy in to that practice in the first place and might be facing a loss of that investment and the remaining partners in the practice were facing a vacancy they had not expected.

The British Medical Association did volunteer to work with the RAMC to organize doctors and do their own recruiting for the military. Initially, the War Office rejected this aid, but this, too, was reconsidered as the war went on and the need was greater. Local plans were developed to deal with the issues that often prevented enlistment, in order to "reassure doctors that adequate health care would be maintained: to ensure that the workload fell equally on those doctors remaining at home; and to formalize the rather ad hoc financial agreements which doctors leaving with the Army had made…"[19] Nevertheless, the continued issues with voluntary recruitment remained. When the British Parliament approved the 1916 Military Service Act, which stated that all men under the age of 41 were liable for military service, the decision was made that medical personnel would still operate under a voluntary scheme, as the enlistment of all doctors would cause massive upheaval in the civilian medical community.

Prior to the US entrance into the war, US physicians and nurses were volunteering their services overseas with groups such as the Red Cross and other nonprofits. One young physician remembered how, when the war broke out, he and his fellow physicians followed events, even though many Americans didn't. "But it was very important to many of the young doctors, including myself. We saw in it an opportunity to practice battlefield surgery, to save lives, and also to see the world."[20] Even if they didn't volunteer, US physicians readily read information about the war that appeared in medical journals, attended talks from people who had served in a medical capacity during the war, and talked among themselves about how things should be organized if the United States did get into the war, which by 1916 was an increasing possibility.

Through these volunteer processes, some became formally attached to military medical units already operating; others worked specifically with the volunteer service they signed up with (this was often the case for nurses); and others were dispatched where needed. Many of the young women who volunteered embarked on this endeavor with a sense of duty and of adventure. In some cases they may have recently graduated from a nursing program and saw the war as a place where their work was most needed at that time. Others saw requests for nurses in local newspapers and persuaded their families to support them. In still other cases, nurses were sponsored by aid societies in the United States to go to Europe and tend the wounded. These young women wrote letters home to their families and/or supporters, describing what they encountered. They also often kept diaries of their journeys to Europe and then of their medical assignments once there.

Some women spearheaded nursing and hospital work through their own force of will. From the beginning of the war, a number of wealthy women from the United States and Great Britain focused on nursing work and equipping hospitals as their contribution to the war effort. This was not an entirely behind-the-scenes effort. Many of these women directly served as nurses, volunteering to go to France and Belgium themselves. Their volunteer efforts were not always appreciated by the military establishment. In the case of Great Britain, they were downright dismissed and derided, being told they were not needed and certainly, as women, were not appreciated. Determined to get to the front as quickly as possible, some even would go around the Red Cross's rules for volunteers and hospital-building. Instead, these women went directly to the French to offer their services in France, and they were heartily welcomed as the French government saw the circumstances of the war as more desperate for themselves and did not turn away aid.

The elite women who pursued this line of work had the money to back their efforts, but they also "possessed 'social capital'" that they could exploit to get their efforts approved.[21] They could purchase their own supplies, arrange for space for their own hospitals or care centers, and recruit more volunteer nurses to work there. These locales also found themselves as the home base for many female doctors who had not been able to get approval to be in the military medical corps. These women also had the standing to go up against the British military and political hierarchy, who in its patriarchal approach had hoped to dissuade their presence.

The work that nurses performed was most appreciated by the patients in their care. It was the nurses with whom the patients had the most continual contact. What can be gleaned from the diaries and letters of nurses working primarily in France demonstrates how much care they were expected to give a variety of patients and, given the realities of how many patients they had to see, how dispassionate they at times had to be. It is clear from reading letters sent home and diaries kept during these years that it was an exhausting and trying time. It is also clear that the nursing staffs did have their "favorites" among the patients.

The volunteer nurses were perceived to be different from the professional nurses. The professional nurses had made nursing their calling, their career. They may have gone through rigorous programs to become a nurse. The volunteer nurses might have among them women who came from the upper class and who had decided to do this work for patriotism, but who planned on going back to their "regular" lives after the war was over.

On the Western Front, women working on medical staffs were almost always serving as nurses. Illustration 3.1 illustrates this phenomenon. On the home front this was not necessarily the case. In Britain the War Office refused to allow female doctors to be sent to the front, but at home "the role of women was to relieve male doctors in the home hospitals."[22] However, the need was so great that other Allied governments were happy to have the assistance of voluntary organizations established by women in the medical profession. In the second half of the war, the British War Office relented and allowed female doctors to be posted to positions in the Mediterranean. A complicating fact for these female doctors was that male doctors were commissioned officers, whereas the women were not, so they always had to defer to the higher-ranked officers, including over issues of proposed treatment.[23] In Britain this continued, and full rank and commission didn't appear for women until 1950.[24]

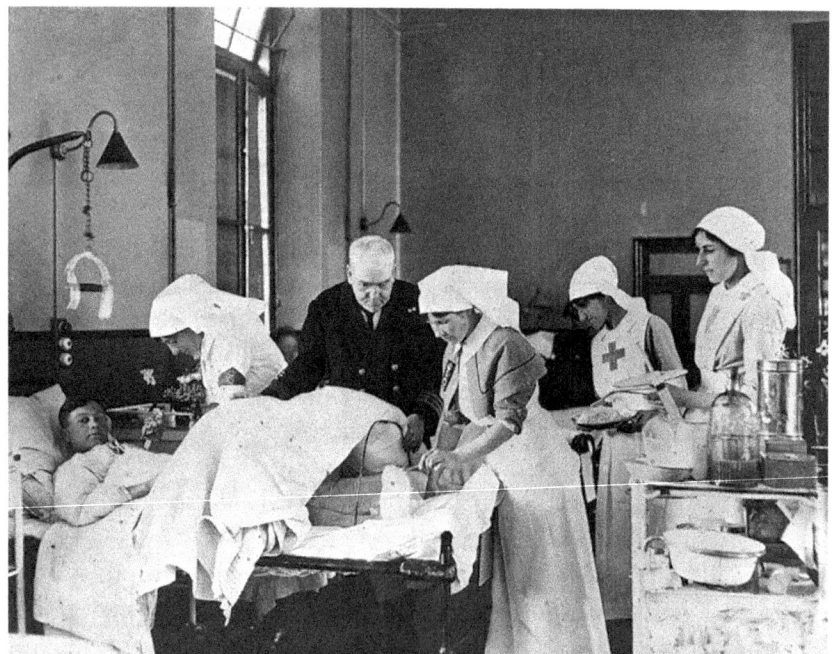

Illustration 3.1 Nurses and a doctor tend to a wounded soldier.

Photo Credit: 'World War I: doctor and nurse treating a wounded soldier'. Credit: Wellcome Collection. CC BY https://wellcomecollection.org/works/gb5peeu7

Male doctors, too, embarked on these volunteer efforts. In 1915, the *Journal of the American Medical Association* brought up the issue of American physicians volunteering for service in the future. One doctor, William L. Hanson, later recalled reading this and being interested enough to write to a British consul about the possibility. Hanson went on to poll his colleagues about it, with some becoming interested enough to form a base hospital unit. Hanson himself continued his studies until the next year, when he was finally contacted by the British consul's office. The British offered to pay his round-trip transport to Europe and back if he was willing to provide one year of service to their troops. He did accept the offer.[25]

Hanson detailed treating German prisoners in London as the first part of his work (he was specializing in ophthalmology) and seeing German zeppelin bombing raids on the city. In his time in France, he continued treating patients and spent time assigned to a field hospital near the front. He noted his great distress one night later in the war, upon realizing that an American division was encamped near the field hospital, and he worried that German attacks on the troops might wipe out the hospital as well.[26]

By the summer of 1916, the American Orthopedic Association was preparing for what it believed would be the US involvement in the war. The association started preparedness groups and orthopedic surgeons were encouraged to volunteer for the war. These volunteer positions were mostly associated with British military hospitals. There was a concerted effort to set up orthopedic centers at said hospitals. The main impediment to this was staff. There were not enough orthopedic surgeons to staff these centers without getting volunteers from the United States.[27]

Post–US Entrance

The volunteer aspect for US citizens became a moot point when the United States entered the war in April 1917. The first group of 20 orthopedic surgeons headed to Britain in May 1917.[28] Although it may seem that such surgeons would naturally continue on to France to practice their craft, it was not necessarily so. H. Winnett Orr, an orthopedist who came with the first batch of surgeons from the United States in May, noted that they weren't needed as much in France as in Britain. He explained that the British had developed an incredibly efficient method of getting the wounded out of France. For example, a soldier might be wounded on one day and be in Britain getting treatment the next. For straightforward, uncomplicated cases, this was certainly a normal occurrence. The problem then became not having these numerous patients neglected at the full hospitals and treatment centers in Britain. Orr observed cases in Britain where patients were healing without the proper splints or with such elements being in the wrong position.[29] These types of problems would mean further corrective surgery for the patient and a longer period of convalescence.

In July 1917, the Surgeon General of the United States ordered the organization of evacuation hospitals for the front lines. Evacuation hospitals had been part of military plans for years, but as the US military had not been in a major war for some time, they had remained constructs on paper only. The first evacuation hospital was organized by the Medical Training Camp out of Fort Oglethorpe, Georgia. The US evacuation hospitals were to be the equivalent of the British casualty clearing stations. They would be near the front lines, and each US Army division would have its own evacuation hospital. They would utilize buildings available in nearby towns (if possible). The United States had seen the benefit of the casualty clearing stations in providing quick patient care. "Speed and efficiency" was to be the slogan for medical treatment for those in need.[30]

To that end, evacuation hospitals were also to be near railroads and outfitted to care for over 400 patients. They would work in conjunction with field hospitals, which provided the first line of care to the wounded. The plan was that field hospitals would resupply themselves from the available stocks at the evacuation hospital. In this way all the equipment used on a patient would travel with the patient through the medical care process, and the field hospital could resupply itself easily in a frontline position.

Each evacuation hospital would have a lieutenant colonel in command along with another 15 medical officers. In addition nearly 30 noncommissioned officers, cooks, and a projected "848 privates of whom 98" would work as ward attendants.[31] It would be expected that surgical treatment at an evacuation hospital would need to be prepared for "a large portion of the seriously wounded… during the strain of violent activity, in which great streams of wounded pour in." Likewise, patients would not see immediate discharge. The expectation was that patients would be in for 7 to 10 days at an evacuation hospital.[32]

The British and the United States both tended to wait when it came to surgical intervention preferring to have the patient in a rear position before attempting surgery. Surgical hospitals near the front, the British believed, were generally failures, though the British did move later in the war to doing more surgeries near the front. However, the French maintained consistently through the war that when possible, they would provide "some sort of surgical equipment in men and material for the wounded man, while still under fire" at forward positions.[33]

As the United States wrapped up its first year in the war, medical officers were apprised of some key points to remember in the organization of the medical service in the war. A memorandum outlined the duties that were essential for the medical service at the front. The first point was that the medical service should take the wounded from the trenches to a battalion aid post giving first aid, and then the wounded should be put in ambulances. Secondly, the medical service should as rapidly as possible get the wounded to a hospital that could offer proper surgical facilities for treatment. The third point was that major surgery would happen at the evacuation hospital, with an eye to avoiding infection.[34] Infection and its concomitant recovery period meant that the military would "lose a part or all of his efficiency as a soldier."[35]

For US medical personnel, there were specific rules and skill sets that everyone should have at the various posts on the evacuation route. As was explained in the use of Thomas splints, "[They] should always be on hand…, and every man should know how to put them on, (knowledge to be acquired by drill), so that under the excitement of battle, or in the darkness of night, the medical department soldier will be able to perform his duty mechanically."[36]

Part of the US view on staffing and equipment at these various positions comes directly from the British and French experiences over the past years of the war. At times there would be an appeal to patriotism pointing out that "the British ambulances are at all times in perfect condition of repair and orderly cleanliness, and some of the best looking machines have had 3 years' service."³⁷ In other words, there would be no excuses accepted for why US ambulances were not similarly cared for and ready for operation.

As described in the previous chapter, specialization became of paramount importance in rendering proper treatment. There was a clear effort to identify and assign doctors to units where their talent would be most useful. The medical bureaucracy grew to facilitate this growth and management. The practice of having special hospitals that dealt solely with orthopedic injuries or solely with brain injuries was an outgrowth of this organizational model. This bureaucracy could be seen in several other areas. Doctors learned to adapt to this bureaucracy, for example, in their embracing of patient charts. Doctors saw the benefit in marking what the patient was suffering from, what they ordered as treatment and when, what the results were, and when the patient was healed. In a system where patients were relocated as their treatment progressed, this record became a necessity.

Stretcher-Bearers

One crucial role in assisting the wounded was performed by stretcher-bearers, so-called because they literally bore stretchers filled with the wounded out of the front lines. They did more than that, however, often administering the very first aid to the wounded and in some periods being the only medical personnel around. Stretcher-bearers were used by all of the Allied armies on the Western Front. Calls for "stretcher-bearer" in the heat of an engagement were a common sound. In fact "when a man was wounded in the advance, no one was supposed to stop to help him; this job was for the first-aid men and stretcher-bearers who followed in the rear."³⁸

Charles H. Horton served as a stretcher-bearer with the Royal Army Medical Corps. He volunteered in 1915 for the service and began his training. In his memoirs he describes the shock of being near the front lines for the first time and how, despite the ever-growing list of new technologies employed by the combatants, he still saw that "these are the days of the horse."³⁹

The danger was most extreme perhaps for the stretcher-bearers as their jobs entailed their presence near the fields of fighting much of the time and venturing into no-man's-land to retrieve the wounded.. Horton describes how, at the Battle of the Somme in 1916, he saw a shell kill a friend and fellow volunteer a few feet away from where he was walking. As his unit made their way through the lines, he saw the parapets of the trenches built up and reinforced with the bodies of dead German soldiers.[40]

Those attending to the wounded during a battle were obviously in the most precarious predicament. But stretcher-bearers tried to make the best of it. Heavy shelling left large craters in the ground, and the stretcher-bearers would put the wounded in these holes and dress their wounds as they could. "It was very seldom that a shell hole would receive another exploding shell, so that is why wounded were placed in the craters."[41] When the shelling stopped, the wounded would be pulled off the field and given better attention. Illustration 3.2 depicts a gathering of wounded who were awaiting transport to other medical facilities.

Illustration 3.2 British Army medical collection point.
Photo Credit: 'World War One: British army collecting point'. Credit: Wellcome Collection. CC BY https://wellcomecollection.org/works/k28gfkyy

Stretcher-bearers might also provide simple remedies for those who fell ill or were injured away from the front. They would provide the first medical assessment of a patient and decide if the patient needed to actually be seen by a doctor. The US Medical Service specified how many stretcher-bearers were needed. "Thirty-two stretcher bearers to a battalion is a bare necessity, and provision should be made to run in an equal number more in time of attack." They further specified that they wanted "strong men, and men of exceptional bravery" because there was "no more exhausting work than stretcher-bearing." These were the frontline medical staff and may be run ragged during an engagement, yet they were to have the highest morale according to the medical administrators.[42] It was a tall order.

Nursing

The fact of life for nurses working on the Western Front is that they were most assuredly not above the fray of war either. Nursing was an essential part of the medical treatment offered on the Western Front, and thousands of nurses served with either volunteer groups or through the military medical services themselves. Nurses were assigned to casualty clearing stations, evacuation hospitals, base hospitals, and rehabilitation locations.

The romanticized version of nursing bore little similarity to the hard reality of nursing. Medical care in the more forward positions came with the risk of bombardment from artillery and the risk of being gassed. Nurses found themselves outfitted with gas masks in the later stages of the war, just as soldiers were.[43] Their days were filled with numerous patients and tasks, and they had little time off. They often did not have enough beds or supplies for all whom they were supposed to take care of.

These women clad in white-aprons were not, as it has been pointed out, a homogenous group. Nurses came from different social classes, including the upper classes. Some had received professionalized training; others had only short courses and had perhaps more enthusiasm than know-how. Different nationalities were represented, just as was the case among the soldiers themselves. They arrived on the Western Front at different times with different perspectives. Volunteer nurses had arrived in France from the United States since the beginning of the war, but suddenly, starting in 1917, approximately 10,000 nurses with the US Army Nurse Corps began to arrive in organized fashion. Tension did exist between the professional nurses and the volunteer ones that was not always easy to overcome in the heightened and stressful environment of war work. The Red Cross offered a series of lectures that were ultimately published in book form that sought to address these tensions:

> it must be understood that no distinction whatever is admissible as to either duties or rights between the paid professional nurses of the hospitals and the voluntary nurses... The same regularity, the same docility, must be exacted from voluntary nurses as from their sisters. And from the professional nurses we have a right to demand the same devotion and the same self-sacrifice as from the volunteers.
> There are not two codes, or two hierarchies... For each is there an equal share of honours, and an equal share of burthens [sic].[44]

Nurses never lacked for work. In the account of her months in France, nurse Sarah Sand Stevenson described the cramped and crowded train journey to the front and the nearly overwhelming duty then of seeing up to 50 seriously wounded men by herself, one other nurse, and two attendants.[45] The daily routine required the re-dressing of the wounds for all the soldiers who needed it, administering medicine to the patients and making sure all the patients had all of their meals. If there were lulls in the day, nurses filled the time by making records and notations on their patients, preparing supplies or getting a quick meal. If a day was busy with new patients arriving, the nurses had to care for them, help decide who needed further care and transport, and find places for those who would be staying at their station or hospital. Edith Appleton, a British nurse, noted one day in 1915 that she went on duty at 5:30 A.M. and did not get off duty until 9 P.M., having no time to eat tea or dinner.[46]

Diaries and letters from those in these positions regularly detail how the war came ever closer. "The first big shell fell quite close to our hospital and the air was so thick with red dust, bits and smoke that we could not see out of our windows... we tried to become use to the five-minutely explosions of big shells close to us, but it was difficult and my knees did shake."[47] Despite their fears and concerns, most nurses tried to remain upbeat in these writings, often stressing how pleased they were to be able to help in a patriotic undertaking.

The customary details of daily goings-on often fail to impart the stress and psychological impact of the job. As one nurse stated, "One lived very many times in a torrent of emotion."[48] Nurses were the frontline caregivers of the wounded. They saw patients who often had multiple injuries that required detailed attention, and additionally they had to cope with supplies that might be in short number with the large influx of patients after an engagement. As Christine Hallett points out in her book *Veiled Warriors*, these nurses might see such an influx of patients in the dead of night, as that was the only time stretcher-bearers might be able to affect rescues.[49] So, after being up all day doing the regular work, they might then be up all night handling these crisis cases. "When the big battles were raging we frequently had sudden orders to clear the wards of as many patients as possible to make room for a big convoy from the front."[50] It was the nurses who provided the hour-to-hour care. They were also needed often in the surgical theaters to help with the instruments and monitoring of the patients.

One account describes a patient where all these realities came to pass. The soldier had an arm broken in several places, and it was still bleeding when the nurse first saw him. He also had abdominal injuries with two wounds in his intestines. Then, on his back, yet another wound that impaired his kidney. Nurses worked tirelessly to change his dressings every two hours, and one of his nurses recalled having to put brown paper under him on the bed as they had no cloth they could use.[51]

Just as medical doctors found themselves treating civilians at times, so did the nurses. This became increasingly true in the last year of the war when mobile warfare returned to the front.[52] Civilians found themselves in the thick of the roving front, and more and more civilian cases were seen by nurses.

Nurses faced all of this work with a level of dedication that was unmatched. They also had to contend with disrespect from male medical professionals who did not value their work, knowledge, or skills fully.

This was still an era in which nursing was fighting for legitimacy, training standards, and respect as it professionalized. It was also subject to the gendered viewpoint that delineated what was woman's work and what was man's work.

Staying the Course

In addition to the physical workload of tending to patients, medical personnel also had to stay abreast of advancements in the medical field. News of newer treatment methods, different things or patterns that were being seen in patients, and what to be watchful for in the future bombarded doctors from a variety of sources. Medical journals such as *The British Medical Journal, The American Medical Association Journal, The Lancet,* and divisional memoranda sent out updates. In some cases these came on a weekly basis. One such memo from the US Army Division's divisional surgeon's office furnished guidance on how to identify gas poisoning and what treatments were available. There were specific instructions for treatment at a dressing station and other instructions for such patients at a field hospital.[53] The increased use of gas late in the war made such communication imperative.

Additionally, there were often meetings and conferences held in France and Britain to address some of these in-the-field developments. Doctors could apply for leave to attend such a conference and often were granted it. Nurses, however, did not have frequent attendance at medical conferences. This is unsurprising considering that in the hierarchy of the medical corps nurses, though seen as absolutely necessary, were generally not afforded the same level of respect as doctors received. Nurses did communicate about what they learned in the war through their own professional journals, which carried numerous such articles, particularly after the war.

Although most medical personnel worked on land, many others served on hospital ships. Hospital ships served as floating infirmaries treating the evacuated wounded. At times they would put into port to pick up supplies and offload patients. In other locales away from the Western Front during the war, surgery was performed on them rather than on land. This was the case in the Dardanelles Campaign in 1915.[54] Hospital ships were usually clearly marked with bright red crosses signifying their humanitarian role in the conflict. For most of the Western Front, however, they served for transport and treatment.

This, though, was not enough. Hospital ships also found themselves the targets of German torpedoes. The German perspective was that these hospital ships were not merely caring for the wounded. "The German Government can no longer suffer that the British Government should forward troops and munitions to the main theatre of war under cover of the Red Cross, and it therefore declares that from now on no enemy hospital ship will be allowed in the sea zone… [if] any enemy hospital ship is encountered, it will be considered as a vessel of war, and it will be attacked without further ceremony."[55] And then after firing on hospital ships, the German government in fact said they were transporting troops to the war zone, so the firing was justified.[56] When Germany announced its policy of unrestricted submarine warfare in January of 1917, the hospital ships were again cemented as viable targets. At the time the Allies vilified this action, and it contributed to the Allied perception that Germany was not fighting in an honorable manner. Such a charge led to the inclusion of the "war-guilt" clause in the Treaty of Versailles in 1919.

Influenza

As the war progressed in its final year, the new calamity of influenza began to take its toll on the soldiers and those involved with medical treatment. This pandemic fed on the constant crowding of people due to the war, whether this was soldiers on a troop ship, military men at the front, or civilians exhibiting their patriotism by attending mass rallies and parades. It further fed on military troops already physically traumatized by years of war and not at their strongest to resist it. The influenza virus of 1918–19 became the world's first modern pandemic. People called it "Spanish flu," which was something of a misnomer. Spain remained neutral in the war and therefore unburdened by newspaper censorship. News bureaus there reported accurately regarding how many in Spain were sick from the flu. Other countries censored that information and provided no real statistics, so it seemed as if Spain was unnaturally hit hard by the flu—hence, the name.

Reports vary as to where the virus first appeared, though Fort Riley, Kansas, seems to be a source of early cases. The initial appearance of the flu was quite commonplace. People were sickened, took to bed, and eventually recovered their strength. The Army Medical Corps in the United States did not show great concern even as it spread from army camp to army camp. Maybe the only thing to truly note at this stage is how fast it spread from person to person. However, as the weeks and months passed, this flu virus mutated into a much more virulent form that began to exact its toll as early as 1917 and continued on its march throughout 1918. Its presence in France also necessitated more action by all Allied authorities. Even before it morphed into its most deadly version, it could incapacitate entire units. "At one French Army post of 1,018 soldiers, 688 were hospitalized and 49 died—5 percent of that population of young men."[57] That 5 percent died was startling, but that nearly 70 percent of the unit was incapacitated could itself be deadly to military success.

The number of cases skyrocketed on both sides of the Atlantic as 1918 wore on. The new flu had become a quickly debilitating disease. A person might feel fine on waking in the morning, have symptoms by mid-morning as a fever spiked, take to bed by the afternoon, and be dead by nightfall. That it was the young and strong who were most greatly affected was surprising and scary. The presentation of symptoms also did not follow the general sequence expected for the flu. It was doctors and nurses at the army camps that began to sound the alarm as the number of cases mounted. In September of 1917 on one day, Camp Devens in Massachusetts reported over 1,500 cases. What they often noted was the development of pneumonia in those afflicted, which seemed to spread like the flu, but again people were hesitant to label it as the flu. Instead, some medical personnel in the United States even speculated that this disease might be a hemorrhagic fever due to the excessive bleeding patients experienced. It was often described as "fulminating pneumonia," but this pneumonia could be fatal in 24–48 hours, which was unheard of.[58] In short, this version of it simply didn't look like the flu. People fell ill so quickly and had such trouble breathing, cyanosis would set it. Cyanosis, the lack of oxygen in the blood, would cause patients to turn bluish in not only their face but their whole body. The lower the oxygen level, the bluer, or darker, they got. Rumors began to spread that this was some sort of Black Death or plague. Doctors themselves continued to be perplexed with this influenza virus.

People feared contracting the flu as there were no proven treatments to combat it, only palliative care. In civilian life people were warned away from crowds. Theaters closed, schools closed—all in an attempt to stop the spread of disease. In military life it was near impossible to avoid crowds. Crowded conditions—whether in a trench or in barracks in camp—were the norm with millions of soldiers fighting this war. The troops therefore experienced a much faster rate of affliction than the general populace, even if they also shared in the benefit of growing immunity. "Soldiers struck down in the first 10 days or 2 weeks died at much higher rates than soldiers in the same camp struck down late in the epidemic..."[59]

Medical personnel were stricken down just as soldiers were. Although the illness was terrible enough on land, it was even worse in an isolated environment such as a troop ship. An account of the conditions on the USS *Leviathan* by the Medical Officer describes this harrowing experience. The *Leviathan* was ferrying troops across the Atlantic in 1918 when it was hit with an estimated 2,000 cases of influenza among those on board. The medical officer described how "Doctors and nurses were stricken by the disease and thus became not only unable to aid but also an added burden to the overworked medical personnel. Every available medical officer, nurse and hospital orderly was utilized to the limit of endurance." The account ends with the note that "the nurses remained until the last sick man was taken off."[60]

What finally stopped the virus was the virus itself. As John M. Barry describes in *The Great Influenza*, viruses have a tendency for future mutations to render them less lethal, which happened in the case of the 1918 virus. It burned itself out. Also at work, however, was the role of the rise in immunity among the population who had exposure to the virus but survived.

What didn't solve the crisis, though, was the work of doctors both military and civilian. The US Army conducted an investigation of the treatment plan of those soldiers and commands infected. What the investigator found was that the only thing that helped matters was isolation of those who were ill. No specific treatment affected the course of the disease.⁶¹ As it faded away, the medical staffs counted themselves lucky as there had been little they could do to stop it.

Home-Front Treatment

Although many of those who took care of patients did so near the front lines or at the very least in France, there were still significant personnel who treated patients in Great Britain. As H. Winnett Orr had noted, the British could evacuate their patients to Britain quite quickly. This served several purposes: firstly, the patient was in more secure surroundings, distant from the war; secondly, it meant that the British and subsequently the American doctors assigned there were able to make use of facilities that already existed instead of needing to erect new infrastructure from scratch.

For the medical personnel assigned to these locations in England, their experience of war was quite different from those in France. The doctors in England faced heavy workloads and often lamented what they could not do or solve for their patients. They found many days often frustrating. Sepsis was a common problem. Dr. Francisco, an orthopedist from Kansas City, Missouri, explained that it was not uncommon to have entire days be spent in the operating room or applying plasters to limbs. He also depicted a great deal of interaction and teaching among the orthopedists in the London area and of observing new techniques of theirs. On the other hand, even if engaged in a common pursuit, people do not always get along. He also observed other surgeons whom he did not like and thought they made a "bum job" of a surgery and did not have a good technique.⁶²

What does come through in the accounts from the individuals on the home front is how the novelty and excitement of being somewhere new and doing this work fades into the reality of the day-to-day workload that only got heavier. After being posted to Aberdeen from London, Dr. Francisco regularly wrote about spending a great deal of time on records and the paperwork on his patients. On July 23, 1917, the entire entry was "Same old grind."⁶³ Interesting cases were noted as were ones that had unexpected outcomes. One patient Francisco operated on successfully subsequently turned out to have gas gangrene, necessitating a second operation to fully amputate the patient's arm. Francisco noted in his diary how hard hit he was by this; the inability to do anything and simply hope that the patient recovered was frustrating.

It was hard work to be sure and could be quite emotional. But the doctors in Great Britain did not have the intense pressure of operating in a war zone, where one could easily become a victim oneself. Additionally, they could and did leave the hospitals for lunch or other meetings or to go to other medical facilities to see patients or procedures. They had days off as well. Dr. Francisco describes arriving in London and spending a couple of days going about the city, having some parties for the unit before beginning work. He later details quite a lot of tennis playing that summer.[64] There were vaudeville shows and burlesque shows to go to, shirts to buy at Selfridges, a trip to the zoo in Regent's Park, and even a party where he met Mrs. Astor and Lady Asquith. There was the purchase of a new Ford automobile, trying it out, and finding a good place to garage it.[65] There were days entirely given over to work, but clearly the lot of the physician in London was markedly different than that of the physician in France.

The treatment of civilians on the home front often took a backseat to the needs of the military wounded. There was a constant tension between the needs of the war and the needs of the civilian population, with hospitals finding themselves short-staffed as personnel went into the war. The government had to balance this situation, though it never fully resolved the tension. For example, during the war a rise in the infant mortality rate in Great Britain was attributed to the fact that so many of the British doctors had gone to France, which created the lack of adequate medical supervision on the home front.[66]

The home front also saw medical officers and personnel who were on leave from the front. Leaves themselves were generally not long. Ten days might be given. But those serving in France and Belgium were likely to go to Great Britain to escape the war as best they could. There they found, as did the home-front-assigned doctors and nurses, that life continued on, and there were hotels to rent rooms from and luncheons to attend and dance halls and music for their entertainment.

Race

The issue of race among the medical staff developed for the American Expeditionary Force was already evident as the United States entered the war in 1917. The US military was a segregated institution conforming to the Jim Crow laws prevalent throughout the South. Though the North had a more integrated society, it was by no means a panacea for race relations. Discrimination and prejudice existed throughout the United States against African Americans. With many important political positions in Washington, DC, held by Southerners, it was not possible to integrate the military, nor was it attempted in these decades. US military units of African American soldiers had been organized during the Civil War and continued to operate in the Western expansion of settlement after the war. These units on the frontier were known as "Buffalo Soldiers" and often white officers commanded them. One such officer in the Spanish-American War was John J. Pershing (who would head up the American Expeditionary Force in World War I), who secured the nickname "Black Jack" due to his command of an African American unit. The United States would not integrate its military until after World War II. In World War I, the US Army organized two combat divisions of African American troops. Approximately, 40,000 soldiers would serve in these divisions. These troops needed medical staff in their divisions, and in the case of doctors, these had to be commissioned officers who were African American.

A large number of African American doctors would be needed for these tens of thousands of troops. There had been an increase in the previous generation in the number of African American physicians in the United States. "Numbering only 900 in 1890, by 1920 the forces of African American doctors had risen to roughly 3,500."[67] During the war, the US Army would not send African American officer candidates to the regular officer training schools because of segregation practices at the time, so two separate schools, or camps, were founded for African American candidates. One was for officers who would be trained for line duty (i.e. combat) and another for the Medical Officer Corps. This was the only time a medical officer training camp (colored) would be created for African American doctors. In all, 104 African American volunteer doctors were commissioned to serve. The training camp also qualified a dozen African American dentists to serve during the war.

These doctors shared in the risks and rewards of treating the wounded, as did those who were not segregated. These divisions operated alongside white units, as did their medical staff. Out of the 104, only one of these doctors was killed, and he died from a "mortar blast that severed his legs while he was caring for wounded troops in the field."[68] These two divisions saw major action in France, suffering numerous casualties, and were involved in the Meuse-Argonne Offensive in the fall of 1918. The medical corps that attended them stayed busy with patients suffering from shrapnel and artillery wounds, bullet wounds, and gas injuries. In some cases once the armistice was in place, the doctors accompanied the wounded back to the United States and continued caring for them at military bases in the United States.

The experience for these troops in France was quite different from their daily lives in the United States not only because they were fighting a war: race relations in France were markedly different, and the prejudice that was so common in the United States was absent. African American soldiers were not restricted in what establishments they could go into or in what entertainments they partook of. It would be a difficult coming-home experience for these men when they had to return to the segregated United States, having seen that another way of living was possible.

Bureaucracy

The large numbers of personnel involved in the various medical corps ensured that there must be a large bureaucratic organization created to manage all of the moving parts and people. Although some saw it as necessary, it didn't mean that those serving with medical units liked the bureaucracy. Many chafed under it, especially the longer they were involved in the war. This is most clearly seen with the British RAMC. As Ian Whitehead notes in his work on doctors and their relationship with the RAMC, doctors "had been prepared to accept the constraints that military discipline had placed on their professional freedom. In peacetime most wished to reassert their professional independence, and to escape from what they regarded as the excessive bureaucracy..."[69]

The numbers swelled as the years passed. When the war began in 1914, the RAMC (coupled with its Territorial Reserves) had less than 20,000 in its ranks. At the time of the Armistice four years later, there were 13,000 officers, with 154,000 in lower ranks.[70] At the beginning of the war, the US Medical Department had 444 physicians. At the end of the war, it had 31,530. Nearly a quarter of all American doctors served in the war.[71]

Medical personnel continued to be deployed to the front even as the end of the war approached. The logistics for US personnel of funneling all supplies across an ocean had still not been fully worked out. Base Hospital No. 52 was organized in the summer of 1918 out of Atlanta, Georgia. The opening staff for this hospital crossed the Atlantic in July and arrived in France in early August. But they too arrived without their needed supplies and were forced to make do with patients already needing care—"the necessity of caring immediately for patients forced the hurried, makeshift equipment of the hospital with cots, bedsacks, etc., expecting same to be replaced later with proper equipment."[72] This was over one year after the United States had entered the war and after it had already seen such errors happen in 1914 and 1915. One of the criticisms of any large bureaucracy is the level of waste that occurs. With so many rules and regulations to adhere to, a missing piece of paperwork could easily sideline the shipment of needed items. As with the fighting of the war itself, logic did not always apply.

Conclusion

The medical personnel who worked with soldiers (and sometimes civilians) during the war did so under severe hardships and workloads. They sought to provide the best care they could even in cases where a positive outcome was unlikely. When working on the war front, they had few breaks and saw a steady stream of patients. In numerous cases their own lives were under threat. Yet people continued to volunteer for this type of assignment. The medical corps of the various armies and the nonprofit organizations proved invaluable as these governments waged war. There would have been no way for the manpower strength to be maintained among the Allied forces if the wounded had not been properly and effectively treated, enabling many to return to service.

In describing the role of convalescent depots, which served as a place for the wounded to recover, the US position was made clear:

> The humanitarian calling of a medical man must always influence every act, and in its application to the war, the part of the medical officers of the Army is that they constitute a salvage corps for men. Patients who reach the stage of convalescence ... should be sent to a convalescent camp, commanded by medical officers, and gradually reconstructed into fighting men ... Every effort should be used to induct the men out of the convalescent frame of mind into the enthusiastic military spirit.[73]

4

Soldiers and the Medical Front

It was the rare soldier who never found himself in need of some sort of medical care during the war. If he was lucky, it might only be a scrape that needed a bandage that would bring him to a dressing station. But this would have been lucky indeed. In battle after battle, the casualty numbers demonstrate the great injuries the Allied armies suffered. Even when there was not an actual battle, the artillery shelling regularly inflicted wounds on the entrenched men. These might be shrapnel wounds, concussions, or all manner of cuts and bruises. Machine gun fire was also a daily occurrence, causing bullet wounds to whatever flesh might be exposed to the enemy. And then during battle, the injuries became more numerous and often more lethal. As such, soldiers became increasingly exposed to the medical front of World War I as the years went on.

Soldiers in the Great War often wrote letters and kept diaries through their experiences on the Western Front. Mail call was regularly scheduled, and soldiers looked forward to receiving messages from home. It was a comfort, a thread to a normal life, but writing about their experiences also helped them mentally and emotionally process how the war was affecting them. However, they were not always free to write to others what was actually happening. Partly this was due to the censorship in place during the war. Letters could be opened and read by military officials so that no information would leak out that would hurt the war effort or alert the enemy of new military schemes. Keeping morale high on the home front was a goal of both military and government officials for all the Allied countries. Soldiers, giving a vivid description of what battle was truly like, risked undermining that morale. But a certain degree of restraint was self-imposed by the letter writers themselves. Soldiers may not have wished their parents or loved ones to read about the grim turn their lives had taken. They may not have wished to alarm or further worry those at home. In one case in 1918, a soldier was expected to die from gas gangrene, yet he merely wrote home, "Dear Mother, you will be pleased to know I am wounded in the left leg and am in hospital."[1] The desire to protect those at home from further stress was great.

Therefore, most of the more accurate accounts left from the soldiers that fought came in the form of volumes written or diaries published after the conflict was long over. There was a spate of such works to come out as the 1920s began—part of the desire to somehow make sense of the insensible war. Unfettered by censors, it is in these works that soldiers hauntingly described the conditions under which they served during these years. These experiences further reinforced the inhumanity of the conflict to those on the home front and to future generations. Volumes continued to arrive from former soldiers well into the 1930s.

Reacting to the Wounded

When the war began and the first volunteers hurriedly signed up for service, they went into the war with a nineteenth-century idea of combat. War was something to be embraced; it was a time when men were able to prove their mettle and their heroism in the face of the enemy. It was a rite of passage, and romantic notions of victory swirled in people's heads. Reality came down hard on these visions. Soldiers needed medical care for a variety of reasons. It could be a chronic illness, a fever, the flu, a twisted ankle, or it could be because of the conditions of the war. Bullet wounds, shrapnel wounds, severe cuts, and concussions all directed soldiers to medical care.

Soldiers detailed in their accounts how they saw others fall and how they reacted to the carnage around them. Perhaps it was someone they were friends with who was wounded. Other times, they didn't know the name of the wounded man but were moved to record the circumstances of what happened to him. On any given day, a number of people around them would be wounded. Sometimes it was the writer himself who was wounded and provided detailed information as to how he dealt with the wound and the medical services he received.

One of the first steps in recovering from an injury was literally to get treatment for it. And this was not always possible. In the trenches during daylight hours, it was often ordered that there was to be no movement in and out of the trenches as it could be seen by the enemy and those moving would be put at risk. This also held for evacuations for injuries. Will Bird, a Canadian soldier, related a series of events in which one sergeant was killed attempting to take a brief look above the trenches, and Bird realized they could not remove his body for the rest of the day until nightfall came. Then, just a few minutes later, another soldier was shot in the face, and Bird risked going for a stretcher-bearer because this second victim was still alive and would bleed out if not treated.[2] In such cases the author does not always provide a resolution for the wounded. The reader may or may not be told what the outcome was for the injured soldier. And in fact, the writer may never know what happened to their fellow soldier.

For soldiers wounded in no-man's-land, they, too, would have to wait until nightfall for evacuation. That, of course, did not mitigate their pain and crying out. It was demoralizing and frustrating for those back in the relative safety of a trench to hear such cries and be able to do little to alleviate their fellow soldier's distress.

Because early treatment was important in ensuring that a wounded soldier could recover, some doctors improvised, given the tight surroundings of the trenches when people were able to be evacuated, in the hope of getting more of the wounded to aid stations. One French doctor devised a "hammock-stretcher" to evacuate the wounded in daylight that allowed evacuations without people having to expose themselves above the trenches. In still other areas, "chair-like litters" were used to move the wounded out to rear positions for medical care.[3]

All soldiers knew that the right kind of injury or "nice" wound might mean that they would be evacuated and perhaps sent back home for good.[4] It might be the only upside to actually getting wounded in battle.

Prevention of Problems

An ounce of prevention is worth a pound of cure. So, one goal for the medical service was to prevent problems, if possible, and to convince soldiers to take the steps needed to try to prevent medical issues from occurring. These guards against potential injury were not always feasible given conditions of battle, but they continued to be a line the medical corps addressed.

One general issue was malnutrition. The food soldiers got was not always the healthiest and at times not even provided in enough quantity to keep people strong. It could be difficult for supplies (including food) to get through areas of heavy bombardment. "I am very healthy and feel better with every day of good feeding. For a few days I was quite weak from the scanty and poor food and could hardly walk around, strong as I am."[5] Other soldiers regularly reported no rations coming up from the rear for a variety of reasons—rear areas had been gassed, traffic problems, heavy shelling, and so on. Soldiers were forced to eat what reserve foods they had and to scour for food, if they needed extra, from what might be farmed nearby.

There was much greater bureaucratic awareness of the need for food to reach the troops because in past wars malnutrition had taken away the ability to fight of many units. The British Army fighting over half a century earlier in the Crimea saw more soldiers treated for scurvy than for war injuries. Based on this, the military bureaucracy redid its program of nutrition, and frontline soldiers were to receive 4,200 calories per day, though substitutions were allowed from the set diet.[6] These substitutions could be bizarre at times and did little to make the individual soldier feel cared for. People complained of receiving huge blocks of cheese or sardines, but no beef, and of being sent hardtack biscuits instead of fresh bread.[7] Although the military was aware of the relationship between food and morale in the men, expediency often won out over quality.

One of the common ailments for soldiers was trench foot. So named because so many soldiers who spent time in the trenches became afflicted with it, it was product of the atrocious conditions in the trenches. There was often standing water in the trenches from the regular rainfall on what had been good farming land. Soldiers' feet never got entirely dry. Even if not in the trenches, mud and moisture were constant companions for the soldiers. This moisture coupled with colder conditions could readily produce trench foot, which was characterized by swelling and pain, causing a soldier to barely be able to hobble around. One soldier, Harry Stinton, described his brush with it:

> Waking up late the next morning, most of us found that our feet were in such a bad way we couldn't stand on them. Some feet and legs were very swollen, and all were terribly sore. When the orderly corporal came round we told him we wished to go sick but that we were unable to walk to the doctor. He told us all to stop as we were, in the blankets. Later he came back with the doctor, who examined all our feet. The worst of the cases were sent to hospital. The rest of us were told to rub our feet with the whale oil and put on clean socks. We could not get our boots on for two or three days and when we finally managed it we could only hobble about for a short distance.[8]

It was not uncommon to see marching soldiers fall out of line to the ground rubbing their feet and despite exhortations by officers, they literally could not continue on.

In a later incident, Harry Stinton was injured with shrapnel in his arm. When he made it to a dressing station, he was thankful for the small niceties shown the wounded patients. "Before starting on me, he [the medical staff attending Stinton] gave me a cup of cocoa and a couple of cigarettes. All the wounded who came in were served the same and I thought it was rather nice."[9]

After initial bandaging at the regimental level, the wounded would be taken to field hospitals. Stinton found that, though wounded, he was still ambulatory and was expected to walk the four miles to the nearest field hospital. The ambulances were full up with worse cases who could not walk. Stinton began his journey but fortunately caught a ride in the cab of an ambulance along the way. The shells continued, and the driver had to swerve constantly and make his way over already very uneven ground.

The ambulance rides could be quite rough on the wounded. No treatment was provided while in an ambulance; instead, it was mere transport of shelves of stretchers with men on them. It was a bouncing roller-coaster-type ride as drivers avoided shells and ditches as best they could. The wounded would have no idea of what actually was transpiring outside, and for those with broken bones, it could be especially harrowing. Red Cross ambulances, which were used extensively in the war, were often driven by volunteers.

It was important for the wounded to be reached and treated quickly (hence the reason so many dove onto battle fields to rescue them). Soldiers might have a survivable wound initially, but if they were left in the field, as some were for upward of 12 hours or more, complications would set in. Exposure, cold nights, and rain all would sap a wounded man of his energy. If a soldier was not reached for a day or more, the likelihood of survival was low. There are more accounts from those who retrieved such wounded than from someone wounded and exposed in such a way themselves.

Logistical issues rather than dangerous conditions could also affect evacuations. When there was a major push or offensive going on, roads filled with men and transport vehicles and all of the supplies needed, going in one direction—toward the action. Getting the wounded out in the opposite direction could prove quite difficult. The roads too were usually not very wide and not very well kept, as automotive vehicles were still new, and wide, paved roads had not been built for them yet. One US soldier, Robert Patterson, described such a circumstance. In October 1918, the American Expeditionary Force (AEF) was involved in its largest operation of the war, the Meuse-Argonne Offensive. The Argonne region was already plagued with a lack of roads to begin with. "What roads there were filled with trucks bringing rations and ammunition, and it was hard to get ambulances up... there were times when the wounded had to wait over 24 hours before ambulances could take them back."[10] A fellow soldier had been wounded in the leg by a bullet, but the wound did not look too bad. However, because of the delay in evacuating him, gangrene set in, and his leg had to be amputated.

Circumstances could change quickly. The wounded might be able to be moved out in one hour and the next hour be told to stay put. One American soldier related a day as he and another soldier attempted to care for a third soldier who had an injured arm. They were fairly near the front, in a wooded area with a quarry. First, one man went out trying to find something to put on the wound, but everyone was out of iodine. Then, later, the second man went out and somehow came back with whiskey, iodine, and gauze, in an attempt to dress the wound themselves, as there were no medical personnel around.[11]

Even in the first dressing stations, tea would be provided to the British soldiers, and usually cigarettes. Warmth was also a key ingredient of treatment. After the stress of an engagement, some comfort and kind words could work wonders in calming the patients and making sure panic did not set in. The British also devised a treatment for shock in these dressing stations that was later referred to as "the British system." It involved placing a small stove underneath the litter that held the patient and then three blankets on him to ensure that he was kept warm as a way of alleviating shock.[12]

French soldiers would be given tags that would be put on a button to denote where they had come from and even where they had stopped for dressings. When brought in on a train to a rear location, "each has been chalked somewhere on his coat with a big B (blessé) or an M (malade)," so the wounded could be sorted quickly.[13] *Blessé* meant "injured" and *malade* meant "sick." The French also developed a mobile ambulance unit to treat the wounded that housed stretchers, tables, an operating room, an autoclave to sterilize instruments, and even an X-ray room.[14] A soldier who arrived at such a facility could receive quick diagnosis and treatment.

Wounded soldiers arriving at a casualty clearing station or later at a hospital found a level of care of comfort (despite their injuries) that they had not had while at the front. The juxtaposition of these two extremes was not lost on the soldiers. However, it is important to remember that even though conditions were more comfortable in a hospital, it too would suffer from shortages and uncomfortable periods. One history of a US base hospital detailed how cold it was in the wards for the wounded soldiers—"all supply of stoves was meager, one for each ward, which with scarcity of coal and character of wood, made the heating of the wards but poor." When more supplies came in later in the winter, the wards were once again warm.[15]

Surgery

Facing the prospect of surgery was difficult for the wounded soldier. He was cast into a world about which he knew little and would have been given scant information about his condition. This lack of information was in part due to the numbers of patients in treatment and the limits of the personnel staff. Surgeons also did not want to alarm patients (if they were conscious) as to what the surgery would entail, for fear of panicking them. In assessing the surgeon–patient relationship during the war, Steven D. Heys pointed out:

> Surgeons today will spend long periods of time with patients before surgery explaining what the operation will entail and then obtain informed consent... Whilst the situation was obviously different almost 100 years ago and in the setting of war, the following is an interesting insight into what surgeons might want to tell the casualty before surgery:... the patient should be warned in a tactful way that he may feel extraordinarily distressed and apprehensive during the operation... but that he should try to lie and breathe quietly, that struggling and excitement will only increase his distress and that the symptoms will pass off...[16]

It was not uncommon for operations to begin with a local anesthetic only (which is what the above statement refers to) and then later for the patient to have a more general anesthetic administered. So, soldiers might go into surgery completely conscious and, obviously, overwhelmed and distressed.

Soldiers in the rear could gauge what action was going on at the front by watching the hospitals and seeing when they prepared for more patients. "The present one here... is evacuating for 500 wounded, who are expected tomorrow. There were scores of ambulances taking away the sick this morning, many of the poor lads traveling on stretchers. I suppose fighting has begun up the line..."[17] Soldiers had no choice but to come to terms with being surrounded by wounded men and still being expected to continue on their own mission. "We were told to rest. Dead and wounded men were all around us; stretcher-bearers carried them past us to the rear ; the guns went on roaring—but I went to sleep!"[18]

When possible, hotels were commandeered for the wounded. In one town, a doctor bragged about getting such a place: "the best only was good enough for the wounded of our battalion."[19] Other troops were treated wherever possible. Along the Marne in the first part of the war, wounded British soldiers were brought to a hastily erected operating tent in a field. The ambulances began to bring in the wounded at night. The soldiers found themselves placed on their stretchers on piles of straw to be taken in one by one to the operating tent. Their wounds were dressed, and they were taken into another field of straw, with blankets below their stretcher and on them. Morphine was given to those in pain. Hot soup, tea, and jam were given to those who were hungry. By the morning, all the wounded had been cared for and either released or put on hospital trains for further treatment. The tent was taken down, and the group moved on with the brigade the next day.[20] This field ambulance was literally in a field this night, but more often wounded soldiers would find themselves in actual buildings.

During the chaos of engagement, it was possible for a solider to lose his identification tag. If he was rescued and unable to communicate, no doctor or nurse might know exactly who was being treated. Sometimes the wounds might be so grievous that even members of the unit might not be sure which of their comrades it was who had been wounded. One US captain remembered how a soldier had been shot in the face, was unrecognizable, and had perished. He went on the word of other members of the unit as to who it was, only finding out over a week later that the man he had reported back to the United States as dead was actually fine, and the solider who had died was someone else.[21]

Soldiers who had limbs amputated faced a number of logistical troubles after surviving such a surgery and recovery in a world where infection was always a possibility. The British, for example, did not have the room or resources to provide rehabilitation aid for everyone who needed it. Many soldiers were sent home after the amputated limb healed. The idea was that these soldiers would then go to "limb fitting centers," where they would receive further treatment and a prosthetic limb. However, the factories responsible for making prosthetic limbs could not keep up with the demand, leaving thousands of soldiers on waiting lists.[22]

Gas

One type of potential injury that soldiers had to face at an increasing probability was injury through a gas attack. As detailed in a previous chapter, gas was used by both the Germans and the Allied sides during the war. It became used as a weapon first in 1915, but by 1917 and 1918, it was regularly used by both sides. Soldiers feared nothing like they feared a gas attack. The insidious silent nature of its approach and attack was unlike the rest of the war—an environment filled with the cacophony of bullets and artillery. Soldiers might hear the whine of the gas shells being fired, but they also might hear nothing. For example, in the winter, gas might freeze in and on the soil at night, and as the day warmed up, the gas would begin to vaporize and find its victims.[23] Soldiers despised the gas masks and respirators they were assigned. The masks were tight and had a very narrow field of vision. An example of various gas masks is seen in Illustration 4.1. They were hot and uncomfortable, and no one wanted to wear one preemptively. However, soldiers grabbed for them and were grateful for them when they encountered gas. They hated the gas because there was no way to fight against it. The goal became only to not succumb to it.

As gas continued to be used from 1915 through the end of the war in 1918, soldiers had many chances to practice how to withstand an attack and also to see the results of them. Some gases killed or debilitated by

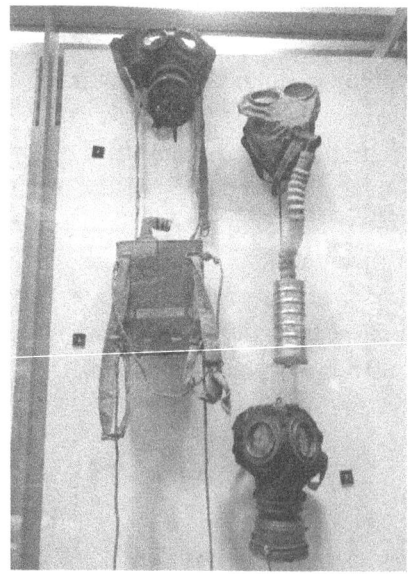

Illustration 4.1 Various styles of gas masks.
Photo credit: Photo taken by the author; National WWI Museum and Memorial

burning the skin or mucous membranes. Others, like phosgene, would cause sputtering and coughing, and then the lungs would fill with fluid that could not be expelled, causing the victim to drown in his own fluids. Gas casualties were heavily enumerated in the base hospital histories written in the United States after the war. For some hospitals, these casualties filled well over a third of all patients seen during their months of operation.

US Captain Robert Patterson was gassed in 1918. He had only a cast-off gas mask at the time that didn't actually filter out the gas, so he threw it away, thinking the gas was not strong enough to injure him. He awoke the next day with his eyes burning and a brown haze over his vision. Taken to the first-aid post, his eyes were bandaged, and he was sent to a field hospital. "I was barely conscious. After considerable bumping about, the ambulance reached a field hospital." He spent nearly a week there. His vision returned after a few days of blindness, but a cough lingered in his lungs for weeks from this brief exposure to gas.[24]

Quick medical care could make a difference in certain gas cases. Soldiers could receive compressed oxygen to help with their treatment. They might receive baths or dousing with water and cleansers to get the gas physically off them. Soldiers might also find themselves losing all their belongings after a gas attack. If the gas got on their packs and uniforms, everything would have to be thrown out since the gas remained on the cloth and could cause further problems if people were exposed to it.

US soldier John Lewis Barkley had such an exposure to gas. He was under fire without a helmet and found one discarded in the woods and put it on. While the engagement was going on, he was so distracted as not to notice, but once things calmed down, he realized his head was tingling and beginning to itch. He sniffed inside the helmet and found that it had been splashed with mustard gas. He tried dousing his head with his canteen water immediately, but there was no effect as his head and neck began to burn more and more. Fearing that he would be sent to the rear, Barkley and friends decided to treat it and some of their other wounds themselves. Barkley ended up having an "iodine shampoo" to clean the remnants of the gas off him.[25] Over the next days, it became a "mass of yellow blisters, like a burn. If one of the blisters broke and the water from it got on the skin anywhere else, there'd be a new blister."[26] He remained completely bald on the top of his head for weeks. After the war ended, he bought a salve from a barber in Germany that Barkley swore made his hair finally grow back thick. He had his full head of hair when he returned home.[27]

The phenomenon of not noticing one was injured at first was a common one. As Fiona Reid noticed in her account of soldiers on the front lines, "many soldiers, especially those wounded in action, did not notice their wounds straight away even if they were severe ... This was a kind of coping mechanism, because on recognizing the actual reality of a wound, a man could be overwhelmed by his fears."[28] The adrenaline that accompanied going over the top or having a shell explode nearby masked the symptoms of the wound for many soldiers. It was only when this adrenaline subsided or when a soldier tried to move again that he realized he was hurt.

Aside from the Allied soldiers and occasional civilians, there is one other group of patients who were given treatment during the war—German prisoners. These prisoners received treatment in the field after they were captured, in the same casualty clearing stations or evacuation hospitals or field hospitals as British and American soldiers were treated. Some were also treated, for example, in Britain after being evacuated due to the nature of their wounds.

German prisoners had all the same types of wounds that the Allied soldiers had. In many of the surviving accounts, there is a tendency of these patients to be treated with condescension by the medical personnel. In one account by a regimental surgeon, he illustrated this in explaining the condition of some German soldiers he came across. They were groaning and pale while under guard in the field. Their clothes had some blood on them, so the doctor checked for wounds and on each man found only pricks on the back. Two of the Germans were "coughing and spitting into their hands and examining what they coughed up." The surgeon then figured out what he believed had happened. A unit of Irish lancers had been through the area and chose to prick the Germans in the back rather than running them through with their lances. The German soldiers feeling the puncture and, thinking it was severe, panicked. "Then the logical German mind told them that this lance-thrust had probably penetrated their lungs; so they spat into their hands to see the blood come up. But of blood there was not so much as one speck, so shallow were their wounds. Then their introspective, literal minds added still further to their terror; for they concluded that they must have internal haemmorhage. Hence their state of mental distress."[29] He found explaining that they were actually alright elicited numerous smiles from them. The stress of battle can certainly affect clear thinking, but the description of the Germans as somehow not being able to figure out they were uninjured on their own would most likely not be given about Allied troops.

Injured German prisoners didn't always come from military engagements. At times soldiers would be sent on volunteer missions to kidnap a German soldier so that he could be interrogated about upcoming German plans. German prisoners could just as easily be shot by their own men as such a mission was carried out.[30] They also might be injured in the kidnapping itself, inadvertently.

If German prisoners were healthy, they might be put to work in some way by the medical corps. Robert Patterson (later Undersecretary of War during World War II) recalled, upon his arrival to France in 1918, seeing German prisoners of war working carrying stretchers of British wounded to ships that would then cross the English Channel back to Britain for further treatment.[31]

Camaraderie

Soldiers wanted to stay with their units, with the men who had become friends. The thought of being sent away unless it was strictly necessary was to be avoided. In US soldier John Lewis Barkley's account of his time in France, he described being cut on his torso by a German bayonet. After running to escape the situation, he and another member of his unit were finally able to assess the wound, which had been bleeding badly. Barkley's friend and fellow soldier applied a first-aid packet and bandage to it. Barkley asked the other man if "he didn't think we'd better see a doctor. 'Hell, no!' he said. 'Don't you know what the pill-rollers'll do? They'll stick a tag on you.'" This meant that Barkley would be sent to the rear for treatment and away from his unit. Despite the injury, Barkley continued with his unit, though a medical officer did assess and rebandage it later on.[32]

In the case of one of Barkley's friends who continued to try to stay with his unit after being injured, he finally found he couldn't. William Floyd had an injury to his arm that Barkley and another had been trying to treat themselves, but once back in camp, it was obvious that Floyd was near to breaking down. He had to be tied down to a stretcher because he so wanted to stay with his unit. The doctor finally convinced him that he needed real treatment and that he would be alright, saying, "You'll find out we're a bunch of fighting doctors too!"[33]

Soldiers who were legitimately ill or wounded did not appreciate those they thought were malingering or were attempting to get out of doing their duty. When ill with a fever and unable to eat, US private Nels Anderson finally went to the doctor and was confined to quarters until better. He wondered why anyone would truly feign sickness to try to get out of the war. "I would rather be well and take my chances."[34] Later he wrote, "Many men are riding the sick already some are lazy but most of them are sick."[35] Such soldiers as Anderson believed that those who faked their sickness were not acting honorably and harmed the military as a whole. There was little sympathy for those who were considered to be cowards.

Soldiers knew that those who were exaggerating symptoms made more work for an already stressed medical staff who were, after all, only human. "I don't think the doctors care much either. I hope they do. Of course good doctors get imposed on by men who want to ride the sick book but that is no reason they should look upon all the men as fakers. I never will go to an army doctor till I am sure it is the last resort then I will have to doubt the interest," Anderson claimed.[36] For all those that exaggerated symptoms, there were countless more stories of soldiers that, though seriously wounded, refused to seek aid as they wanted to continue in their position or maneuver and were forced by their fellow soldiers to finally get medical attention.

Soldiers generally had positive interactions with the medical staff. They appreciated the effort made on behalf of the troops. Still, soldiers at times would try to find humor in their interactions with the medical corps. One battalion surgeon was described as "very literal and gullible," and the men in his unit would delight in "inventing diseases and describing their symptoms to him."[37]

Even the soldiers were impressed with what the medical corps could do with their wounds. John Lewis Barkley described a major over his unit who had been wounded. "He'd been shot over the heart—straight through the front and out the back. But they had done a great operation on him and he'd recovered entirely. That was luck for the outfit." Another wounded soldier returned briefly at the close of the war to see his unit. "He unbuttoned his blouse. His whole chest was punctured with machine-gun bullets. His lungs were punctured. I couldn't see how he'd lived through it."[38] Clearly, the medical personnel who worked on these men were able to save them when, in an earlier war, these troops would have been lost.

And lastly, sometimes the patients were the medical staff themselves rather than the frontline troops. One surgeon described the experience both he and other soldiers had after being under intense shelling. He found himself analyzing the fatigue that he and the others felt:

> We would always want to sleep; anywhere and at all hours: at our meals especially. Conversations were broken off by one or the other falling asleep.... We would wander in our talk and strange delusions filled our minds. They tell me that I talked of horses with white fetlocks; and I am not a horsey man.... After a few days these delusions would fade away; but the excessive sleepiness and a strange appetite for sugar... would remain with us always.[39]

Aside from the cases of medical personnel being injured in combat near the front, medical personnel also succumbed to being patients and victims of the influenza epidemic in 1918. Tending to so many flu patients, the medical staff had near constant exposure to the virus.

There too was the omnipresent reality that death hovered over all of the soldiers. There was the obvious death of a comrade in arms that could happen under fire, but even in peaceable times, death was present. Soldiers had to do burial details, a "nasty job."[40] They had to march through areas that had been shelled, with the dead still there. "We left the dead man. He had several other dead men there to keep him company... It seems hard to leave the boys unburied but its war."[41] The reality was that as often as surviving troops were patched up, it meant the war could continue for another day.

Soldiers and Noncombat Injuries

Despite the gravity of many of the wounds inflicted, other issues troops had to contend with were more annoyance than injury. One US soldier, Nels Anderson, was in France in the summer of 1918, preparing for what would be the Meuse-Argonne Offensive. "I have been itching a great deal of late. I can't find anything on me so it must be a skin disease. Down at the bathhouse today I notice other fellows covered with little red bumps such as I have got. They said they were fleas and would jump off when a fellow took a bath or changed clothes and then jump back on him."[42]

Stomach issues also plagued troops who were contending with the less than hygienic conditions of the front lines. Cramps, loss of appetite, and dysentery were common. One soldier opined that "it took more courage to face the chow line than the enemy."[43] Issues with the food supply further contributed to these ailments. The toll that the front lines took on the soldiers and the stress of the warfare further compromised their bodies, breaking down their immune systems. Lice, rashes, constant irritations then could prove to be more dangerous if infections set in through open sores. There were procedures in place to minimize the skin irritants. Soldiers might be given hot baths and have to turn in all their old clothes. They would receive new outfits when they emerged from the baths to rid them of the lice, what some called "cooties." This didn't permanently solve the problem as the "cooties" would soon turn up again, but it did provide a measure of relief. One US Army officer always insisted on his soldiers having clean underwear, saying, "Keep your underclothes clean… It may save you from infection when you get plugged."[44]

After being in the field and under fire continuously for several days on various assignments, a US soldier related being taken to be cleaned up. "They gave us medicated baths and deloused our uniforms. After the bath we smelled like a dog hospital. And when our uniforms came back there was so little of them left that some of us had to wear blankets to piece them out." These men were later given fully new uniforms and equipment.[45]

Base hospital histories demonstrate the wide variety of medical sufferings that brought soldiers to them. To spend time in a base hospital meant that the soldier had already been treated at first-aid and casualty clearing stations, and their conditions needed further care. In addition to the expected cases of gas and flu, one history detailed the numerous cases of tonsillitis, tuberculosis, pneumonia, gastro-enteritis, rheumatism, and even migraines bad enough to confine a soldier to a base hospital.[46] Most surgeries involved gunshot wounds, but there were also surgeries for hemorrhoids, appendicitis cases, hernias, and even varicose veins.[47]

The same circumstances that exacted a toll from more benign ailments also fostered the spread of contagious diseases such as typhoid fever and, later in the war, influenza. Life on the front or in the crowded barracks held the same environment for spreading disease. Crowding was especially a problem in training camps and on transport ships. A US officer gave an account of his crossing in the spring of 1918:

> The cabins for us officers were comfortable, but the quarters for the soldiers down below were fearful. This was my first sight of the hardships of war. The overcrowding was beyond belief. Rough tables had been knocked together and placed in the lower decks. The men slept on these tables and under them on the floor, and as many as possible slept in hammocks. There were no bunks of any sort… The air down below could be cut with a knife.[48]

On another ship on another crossing, a Canadian officer gave a similar report. "The quarters for the officers were tolerable, but those of the men were vile, crowded and filthy. I do not think that I have ever seen so many cockroaches in one place anywhere before or since."[49]

Another soldier described similar overcrowding in the transport trains once in France:

> The crowding of soldiers into the cars beat anything I had ever seen. Each freight car was supposed to hold 40 men, which was fitting them in very tight; but in fact most cars held 50. None could lie down, and many could not even sit down on the floor. The choice places were in the doors, where men could sit and hang their legs out… we were on the train 3 days.[50]

John Lewis Barkley, a US soldier and Congressional Medal of Honor winner, gave a vivid description of what disease in these types of close quarters could look like:

> Before I knew what was happening to me I'd been dumped into a pest camp with a nice case of measles. My tentmate had mumps, so I took that on too. There were so many of us sick that we got mighty little attention ... Finally a major doctor came around to look us over. Sick as I was, I could see that he was excited. And when they went on to the next tent, and found one of the boys in there dead, he went wild.[51]

Crowding and overcrowding was a fact of war, but it also substantially affected the rates of communicable diseases and the illnesses of soldiers fighting the war. When one remembers how many men were packed together on both sides on the Western Front in what was a very concentrated space, it is perhaps a surprise that there weren't more disease outbreaks. But the fact that the overcrowding could not be alleviated was a complication the medical corps had to learn to deal with. There were a number of communicable diseases seen among the military, including meningitis and measles. The overcrowding also proved to be a serious problem when the influenza epidemic appeared, which spread quite quickly from person to person.

The focus on hygiene also was a daily reminder from the military itself, not the medical corps alone. The US military further particularly focused on the eradication of venereal disease among soldiers describing cases of it "more serious" than the wounded soldier because this type of "disability was preventable and in acquiring it he [the soldier] did not register a blow against the enemy in front but literally gave a victory to the enemy behind the lines."[52] All the armies had suffered when soldiers were not fit to be sent into combat due to venereal disease, so as it entered the war, the United States hoped to impress on its soldiers and draftees how important it was to prevent these conditions. The military warned that the danger came not only from red-light districts but also from prostitutes or camp followers who might appear more benign. Combatting vice, therefore, as it turns out, also came under the purview of military medicine.

Although it was a rare and lucky soldier who never needed any medical attention during the war, it was still an aspect of the war that soldiers might be involved with in other ways. All of the hospitals, for example, had to be built by someone, and those people turned out to be soldiers. Numerous military units were assigned to such construction projects throughout the war.

H. H. Storm was assigned to an engineering company and tasked with hospital construction in 1917. He was excited about the project and reported, "The job we have here is one of the finest in France—building a hospital for wounded. I am very glad I have work of constructive character, not work of destruction... In comparison with the rest of the American soldiers we have seen about us, we are the very best and most industrious of them all."[53] The companies building the hospital would race against one another to complete their buildings. Additionally, water sources had to be found (or built), and pipe had to be laid before the buildings themselves could be erected. Some parts of the hospital (such as the barracks) arrived in sections by railcar, and they only needed to be transported to the site and assembled to be ready. Streamline efficiencies such as this helped speed the construction.

Hospitals were often composed of numerous buildings that would become home to thousands of people. Therefore, the facilities had to be sufficient to take care of the needs of so many. This meant having enough barracks and beds, but it also meant having enough latrines. One account of a hospital describes the obvious: "The two bucket latrines provided for patients, 30 seats for 1600 patients, proved inadequate." Construction workers had to blast out solid rock in the area to build new latrine pits.[54] Such projects at a hospital could continue to occur and occupy workers for months after initial construction began.

Aftercare

Soldiers could have continued engagement with the medical front, depending on their type of injury. For those who would be receiving plastic surgery because of a wound, often multiple surgeries were needed over a year or more. A patient in these cases, though, might not need to be in a hospital the entire time and would simply come in for appointments, assessments, and the surgery itself. Other types of injuries, such as orthopedic ones, required much more daily follow-up. To that end, the

British began an orthopedic reconstruction hospital that would address the fact that this was specialized care that could not merely be taken over by the existing hospital system. Doctors argued that the government bore responsibility not only "for treating wounds in the acute stage, but also the adequate medical, vocational and social care of the crippled soldier."[55] Contending that such orthopedic hospitals also needed to include "the workshop, the agricultural school and the business college," the idea was that through various physical therapies and vocational therapies such veterans would be able to be productive members of society.[56]

Aftercare also involved making sure that prosthetics were available for the soldier who had lost a limb, whether an arm or a leg. After amputation and immediate healing, a soldier would be given a provisional prosthetic that could last up to a year. Over time, the remaining stump of the amputated limb would shrink in size as swelling from the trauma subsided and some of the muscle tissue atrophied, and the provisional prosthetic could be adjusted for these sizing changes. Soldiers received physical therapy with their provisional limb before they were discharged from a medical facility. Certain hospitals were outfitted with orthopedic surgeons in the United States under the auspices of the US military. Some would receive hundreds of soldiers who were originally injured overseas. Others received comparatively few cases and ended up putting their orthopedic surgeons on general surgery rotations instead.

For the wounded soldier, once a stump had finished shrinking, a permanent prosthetic would be provided. In most all cases, this occurred after the patient had been transported to his home country. In the United States, this was the responsibility of the War Risk Bureau. The sockets that would hold the stump (either for a leg or arm) were made of wood and were specially routed and fitted precisely, so this had to be done after all swelling of a stump had ceased.[57] Due to the specialized nature of building a prosthetic, this was a significant expense.

It was the duty of the governments and their militaries to ensure a continuity of care for these veterans. In June of 1918 in the United States, the Vocational Rehabilitation Law was approved. It outlined that it was the "complete responsibility of the Medical Departments of the Army and Navy for all measures aimed at the physical and functional restoration of sick and wounded soldiers and sailors." It further specified, "There is to be no interference with the jurisdiction of the military medical authorities" when it came to these issues.[58]

Hospitals in either the rear or in home countries often had soldiers as patients for prolonged periods. Life in the hospital could be monotonous for the patients who were spending weeks in such an environment. There were such varying limits of physical mobility that one of the best things that could be done for these patients was to engage their brains. Autograph books with notes and jokes from their friends were popular as they could be taken home when a soldier was released from care. The image pictured is of one a staff member at a Red Cross hospital kept (Illustration 4.2). Many hospitals began their own local newspapers that were written, edited, and published by the patients staying there. The benefits and therapy of such magazines and newspapers were immediately obvious. This started quite early in the war, as Jeffrey S. Reznick has noted, in Great Britain; "by the end of the first year of the war, nearly every general military hospital in the country had initiated a similar literary project. Throughout this period, auxiliary hospitals that had sufficient funds and staff also began to publish their own magazines."[59]

Coming out every week or every month, these hospital newspapers provided entertainment and engagement for the patients. They advertised

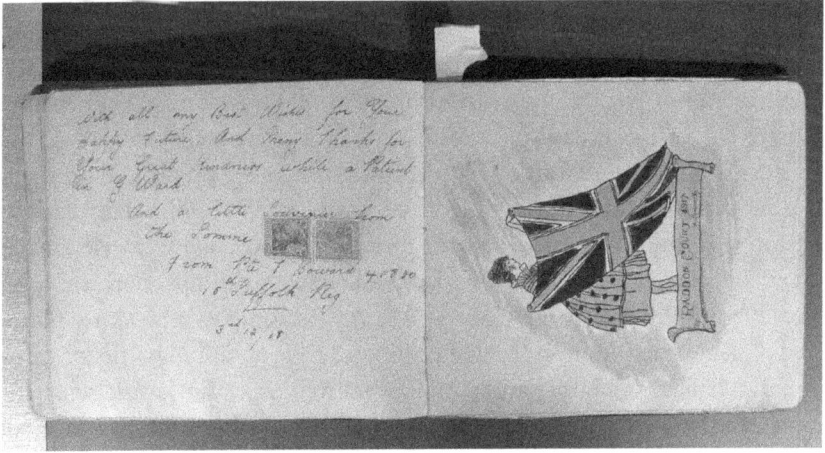

Illustration 4.2   Autograph book from a hospital.
Photo credit: Europeana Collection, CC0

different evening and daytime activities, featured jokes and poetry along with various games, and generally patients poked fun at each other and the staff. The newspapers also offered news about the war when it was still going on and later on about demobilization timelines. They brought some levity and information to patients who often felt cut off from the rest of the world and their families.

Psychological Care

For some soldiers medical care took the form of psychological care as their minds tried to make sense of carnage, shelling, loss of friends, and the devastation all around them. This was not a new phenomenon for military troops. In the US Civil War, the psychological trauma of war was called soldier's heart. In the Great War, doctors came to call this trauma shell shock. In World War II, it was called battle fatigue. Then, in the late twentieth century, it took on the name post-traumatic stress disorder (PTSD). Shell shock was an apt name in World War I, because most doctors directly attributed the problem to the intense and prolonged artillery shelling patients had survived. Because they had not seen such a large number of patients suffering from such in the past, they thought it must be due to the new realities of battle and that the conditions were in some way caused by the reverberations and concussions of the artillery. They were not wrong in this assessment, though they misattributed the cause. The constant exposure to the type of warfare seen in WWI did in fact cause the distress. But obviously, it was not solely predicated on shelling, as some doctors began to realize as early as 1915. The daily actions of the war, the constant stress and threat of death, having to view what happened to fellow soldiers, and being asked to continue to perform one's duties as a soldier were enough to push any person past a breaking point.

Soldiers suffering from this condition did have an uphill battle for treatment. Medical personnel and psychiatrists were not always sure what the best treatment of such patients was, even if they were convinced treatment was necessary. Some psychiatrists believed only quiet rest and a calm environment could solve the soldiers' problems. In Great Britain, there was some practice of psychotherapy as a method of treating the patient, but this was not regularly mandated. Instead, in many instances, those suffering from shell shock were seen as hysterics who needed firm guidance and even harsh punishment to "force" them to behave bravely. Shell-shocked soldiers would be mocked, called cowards, and even given electric shock therapy in an attempt to "cure" them of the problem. Cowardice was not accepted, and being a labeled a coward was downright dangerous. All the Allied armies had cases where the authorities decided to execute a soldier accused of cowardice to provide a lesson to others. More sympathy might be shown for the soldier who began to exhibit such symptoms after being involved in a great deal of combat, but soldiers who developed shell shock away from the front lines were treated more harshly.

The symptoms of shell shock or PTSD do not disappear once the patient is removed from the battle, however. Patients suffering from this condition did so in some form for the rest of their lives. Psychologists and psychiatrists studied patients who had the worst exhibition of symptoms in live-in facilities both during and after the war. These patients were unable to cope with the realities of everyday life. Some abhorred loud noises. Some developed crippling anxiety. Others suffered from extreme abdominal pains that did not go away. Nightmares and tortured sleep were common. In the worst cases, the mind retreated into itself, and the patient could not have a meaningful exchange with another person.

Even if not suffering from full-blown shell shock, soldiers often attributed odd physical quirks or mannerisms to their exposure to intense shelling. US soldier John Lewis Barkley suffered from a stutter for his entire life until serving in France. An artillery shell blew up an inch from his hand and knocked him out. When he regained consciousness, he no longer stuttered when he spoke, much to his amazement, and he credited this change to the shelling.[60]

Soldiers developed rationale and explanations as to how to keep safe during an exchange of fire or shells. These thoughts were as much to give them comfort as anything else. One US private recorded in his diary, "I felt a tense feeling. ... but it wasn't fear. ... I don't believe one ever hears the shell that hits him because it has to hit him before it makes a noise and I noticed that I could see the flash of the explosions before I heard it and whenever I hear a shell hissing or whistling I know it is not going to hit me."[61] There was rarely a clear reason why one soldier might get hit and another not: why one unit was lucky enough to be in muddy ground while being shelled, and the shells threw up mud but little shrapnel, whereas another unit was on a hard surface and would suffer many casualties. These unanswerable questions contributed to the psychological toll on soldiers during the war and the feelings of guilt many had for surviving the war.

Conclusion

In Captain Robert Patterson's memoir, he noted how, after the armistice went into effect in 1918, "many of our original men who had been wounded returned from hospitals." He further detailed how many who had been injured and held as prisoners by the Germans similarly began to come back through the Allied lines and rejoined their units. By the time everyone was accounted for, his company stood, ironically, at 20 percent above full strength.[62] Here, at the end of the war, despite all the hardships, in some cases enough men had recovered to fill the vacancies. That this was even possible is a testament to the work of the medical personnel and the strength of the soldiers as they embarked on their recovery programs.

The medical innovations changed the lives and prognosis for these soldiers arriving for medical care. New surgical techniques meant their bodies could be repaired in a way previously impossible. But it was the full complement of changes that meant more soldiers survived their time at war. "Progress was made in anaesthesia, resuscitation, blood transfusion, antisepsis and radiology which all contributed to constantly improving survival rates."[63]

Soldiers in the war suffered all types of injuries necessitating medical care. They relied on each other and the medical corps to provide them treatment. In general, the medical care they received was vastly improved over what soldiers received in previous wars both as a product of improved science and medical knowledge and through the improved organization of medical corps and continuity of care in this war when compared to previous ones. Accounts of soldiers such as one who was "hit by three bullets in the arm and one in the chest" ended not with demise, but instead a follow-up that they were treated and were then out of the war.[64] This in itself was an amazing outcome and one that soldiers came to rely on—that as much as the war chewed up soldiers, if one was wounded, he would receive the best medical attention possible and the best chance for survival.

5

Effects of the Medical Front in the Great War

As the war wrapped up, doctors and other caregivers turned their time and thoughts to other concerns than the daily grind of work. The outbreak of influenza was ratcheting up and certainly occupied the focus of many in these waning months of the conflict; it would continue to do so into the early part of 1919. Some looked forward to demobilization after the armistice, hoping that the peace talks in 1919 would bring about a period of harmony after these long years of fighting. Soldiers excitedly looked forward to returning to their families and homes. Others approached the future with questions arising out of their time on the front. How could the new medical practices honed during the war be taken back to the civilian side of medicine? The success of the civilian medical endeavors was not simply proving the merit of the medical innovations and knowledge from the war; it also involved often personnel effort by individual physicians to spread that learning themselves.

That medical practice in general would benefit from the knowledge and learning achieved on the medical front was a given. This was true with all war. With so many opportunities to hone their craft, doctors emerged from war knowing better treatment methods. In an article for the *British Medical Journal* right at the end of the war, Professor James H. Nicoll noted, "A large measure of revision of our standard books, however, will be necessary in the light of the experience of the war."[1]

During the four years, there had been regular allusions as to how the practice of medicine and treatment would change. The new practices and techniques were made available to the civilian medical community as soon as they were developed and proved valuable. So, civilian doctors could be reading about them only months after they were first utilized in the war. The *British Medical Journal*, *The Lancet*, and the *Journal of the American Medical Association* all ran articles written by wartime doctors. "The medical need to communicate both the short-term and long-term consequences of injuries and the substantial clinical experience gained by this new generation of military doctors could not have been gained in civilian practice."[2]

Doctors who had served in the military medical services also looked forward to their own demobilization. This happened quite quickly in the case of the British Royal Army Medical Corps, with some fearing that the rapid demobilization meant that existing military patients would be left without care.[3] This created an obvious hardship. However, the demobilization of these doctors also meant that they would be in civilian practice again quickly, continuing the dissemination of the war's medical knowledge to their home countries. This meant that these new practices were diffused "more widely to other countries than would have happened under normal conditions."[4]

In the case of American surgeon Harvey Cushing, when he heard the armistice might soon be at hand, he began speculating how the war would wind down in terms of the wounded. "The first thing necessary is to beat it home and get a gigantic triage started at the Staten Island plant where cases can be combed out, sorted, and routed to the proper hospitals."[5] Though he also surmised it would be quite difficult to organize and persuade government officials to approve of this plan. In the end, Cushing and other medical personnel spent several months after the armistice continuing to tend the patients in France and preparing paperwork to facilitate their transfers to the United States. Cushing sailed home on a troop ship in February 1919, noting that tens of thousands of soldiers were still stationed in France.

Cushing himself is a good example of what this war meant to medicine. Harvey Cushing had already been a practicing surgeon for years when the war arrived. He was in his late forties and had worked at Johns Hopkins

Hospital and was a professor of surgery at Harvard Medical School, specializing in neurosurgery and brain surgery before World War I. He was already widely respected on both sides of the Atlantic, with a secure reputation as an innovative and talented surgeon. He first traveled to France in 1915 with a volunteer ambulance unit organized by Americans. He was unsure as to how much good such a group of volunteers could actually do.[6] However, he found himself dealing with a number of interesting cases, though it was rare to have the ambulance unit (in effect, it was a hospital he was working at) overwhelmed. The same spring he toured the British military medical services, and he found himself surprised at the number of cases with which the British were forced to contend. He even expressed a kind of envy of how many neurological surgeries his counterparts had the opportunity to handle and the possibility of advancing medical knowledge.[7] Cushing and fellow medical volunteers did tour the front lines and saw the great destruction and the cases of soldiers who were too wounded to be moved. Such tours and cross-exposure between doctors from various countries was regularly done. Doctors coming into the war zone wanted to know what was going on, and the doctors practicing there were excited to share information about their cases with newcomers. Cushing returned to the United States by the summer of that year. He remained affected by the war, but he resumed his work in New England and continued his surgeries focusing on brain tumors.

Because of his training and his inclinations, Cushing was selected to head up a surgical unit in France when the United States entered the war. Cushing performed surgeries throughout the remainder of the war, as well as facilitating medical conferences and meetings of physicians in Britain and France. After the war, he returned to teaching and his practice, becoming known by many as the "father of neurosurgery." The war gave him too the opportunity to experiment with different techniques. He developed a method of using magnets to remove metal shrapnel from wounded patients. He continued to experiment and research this treatment method in the 1920s and 1930s.

The war had seen medical personnel not only develop new treatments and approaches but also adapt to the regular use of new equipment in diagnoses and treatment. Doctors found novel ways to use equipment to better treat patients. The X-ray machine had proved invaluable during the war to pinpoint the trajectory and location of bullets and shrapnel, ensuring more clean extractions. The senseless probing, attempting to find the pieces of metal (and the commensurate injuries from such) were done away with by virtue of utilizing X-rays. The use and reliance on these machines, though, did not happen immediately. There was a learning curve for all medicine in this way, and surgeons used to their previous methods of bullet extraction were not keen to adopt the X-ray machine initially. Individual technicians such as Marie Curie's own daughter, Irene, tackled this work in the fall of 1914 and moved to train other women on how to be X-ray technicians, thereby opening another avenue for women to serve in the medical front. Eventually, 150 female technicians were dispersed to X-ray units.[8]

But the X-ray machine could help patients in other ways too. Doctors found a use for it in prepping plastic surgery patients. When tissue, particularly around the eyes, ears, and mouth, became unmalleable and needed to be maneuvered for reconstruction, doctors turned to sessions of X-ray exposure. "This tissue is rendered considerably softer by from one to three intensive treatments."[9] X-rays were also deployed on skin to be used in skin grafts to prevent hair growth in the newly positioned skin. Although the lack of hair was not permanent, it could help to more normalize appearance until the wound was better healed. As skin grafts were regularly used with injuries to the face and neck, helping patients navigate their reentrance to society was as much psychological as physical.

Veterans and Medical Care after the War

Although many veterans of the Great War had no or few lasting physical effects from their injuries, others faced an uncertain future. In a time where disability was shunned and looked down on, soldiers were encouraged to proceed in their lives as if they had no physical impairment. In the summer of 1918, as the war was entering its last months, a new magazine made its appearance in the United States. *Carry On: A Magazine on the Reconstruction of Disabled Soldiers and Sailors* spoke to the reality thousands of US soldiers were already facing just over a year after the country's entrance into the war—a need for rehabilitation and a new approach to the future.

Articles detailed how the government would compensate the "handicapped fighter."[10] Financial amounts varied, depending on the severity of the injury and the number of dependents the former soldier had to support. Other articles discussed job opportunities for the wounded. A great number of them were physically demanding outdoor jobs such as farming. In one caption, a perhaps overly optimistic description accompanies the photo of a veteran who has lost his left arm and is standing examining a plant in a greenhouse. The author extols that "in all parts of America there are splendid opportunities in farming and gardening, where loss of limbs is no drawback."[11]

By design the style of the articles is to uplift the reader and encourage those wounded in the war that there still was a bright future for them. It might be difficult for reality to meet up successfully with the optimism. This was still an era before the overall mechanization of agriculture that was more fully felt by the 1930s. The farming career that was so often mentioned in the magazine was still a laborious job. Conscious or not, careers in farming would move these disabled veterans out of the urban centers where they may be thought of as a more visible "problem" and instead put them in a situation where they would be expected to provide their own food for themselves, thereby being more self-sufficient and needing less government funding.

The lessened government funding was a clear goal of the rehabilitation efforts. And these efforts were a prime example of how the new know-how from the war directly encountered the civilian world postwar. Amputee soldiers represented 5 percent of those US soldiers wounded in the war.[12] As Beth Linker details in her book *War's Waste: Rehabilitation in World War I*, the US government hoped that rehabilitative efforts would lessen the cost of government pensions to the wounded for this war. After previous wars, officials made the assumption that those who had been severely wounded would not be able to provide for themselves in the future; hence, the long-term pensions. Such costs for past conflicts had reached astronomical heights.

A distinct path to avoid pensions was to ensure that when a soldier was wounded, he received proper treatment and would be functional for work after the war. So, if a soldier had an injury to a limb, splints would be applied (as detailed in Chapter 2), and these splints would continue to be applied in an "assembly line of care," as Joel Goldthwaite, chief of military orthopedics for the American Expeditionary Force, described it.[13]

The medical front then continued from treatment on the battlefield to treatment on the home front, as "reconstruction hospitals" would be used by the US Medical Department to provide rehabilitation and treatment to those who had been wounded.[14] Hospitals such as Walter Reed in Washington, DC, focused on the aftercare of these soldiers. At Walter Reed, the Limb Lab was created to introduce prosthetics to those who had received amputations. These prosthetics were as much about obscuring the amputation as they were about functionality. Lifelike prosthetics such as the E-Z-Leg were preferred by orthopedists, even though many of the patients found the model supremely uncomfortable. The focus was always to make these men appear able-bodied. It was, in a sense, to erase their mutilation and injury from having occurred. News articles regularly detailed how amazing the work was that meant these men could walk, bicycle, and move their prosthetic arms and legs as if they were the natural versions they had been born with.[15] This glossing over of what was real trauma to the bodies and minds of these men did little to help prepare them for the next stage of their lives.

Not all patients were pleased with the aftercare treatment they received. And there were times that those tasked with taking care of veterans failed to do so properly. Some soldiers continued to petition for care for years after the war ended, soliciting documentation from fellow former soldiers attesting to the nature of the wounds and that they were obtained in battle. One officer wrote to the Veterans Bureau about such a case for another soldier, describing where the wound was, how serious it was, and that he had personally bandaged the wounded man, and that he could not understand why the Veterans Bureau had completely failed to provide medical care for this veteran, as it was clearly a wound received in action.[16] In following British wounded who returned to England, reports indicated that over 50 percent of those who had suffered serious battle casualties had chronic conditions of the bones, joints, muscles, and nerves.[17] None of these types of maladies could be easily resolved, and they required years of continued care.

"Although standardized rehabilitative care had a military birthplace, it quickly became part of the expected civilian practice in nonmilitary general hospitals."[18] The surgeon general of the United States detailed some of the responsibilities of these aftercare programs. They included "reeducation, or the vocational training of the disabled, including the blind and the deaf; rehabilitation or the social adjustment of the disabled to suitable employment, and the education of the public to a proper attitude toward the crippled soldier."[19] The American Red Cross also began a campaign for public awareness and fair treatment, declaring that although the government would provide the best of medical care, whether the injured veteran succeeded again at life would depend on "whether the attitude of the public operates as a help or hindrance."[20]

Unfortunately, in a number of cases the public sought to relegate not just the war itself, but its wounded soldiers, to the background. As Jeffrey S. Reznick has noted, in Britain "disabled heroes ultimately found themselves ... swept out of the post-war labour market."[21] And this was despite the focus on prosthetics and rehabilitation giving these men a semblance of being able-bodied.

Another group of veterans who needed and often sought some sort of care after the war were those suffering from psychological trauma because of the war, the shell-shock patients. In Great Britain the Ex-Services Welfare Society, or ESWS, provided a sympathetic portrayal of these former soldiers. The ESWS was the only charity "designed specifically and exclusively to cater for veterans suffering from psychological injury."[22] There were limited options for those suffering after the war, as the government was slow to deal with these cases. Asylums were often poorly maintained and often were a place to house people rather than to treat them. The ESWS focused on providing an alternative to the asylum for those former soldiers who could no longer live with their families because of the nature of their psychological conditions. Demobilization, instead of being a joyful time, often brought more stress to the former soldier. The veteran could not cope with the realities of civilian life, and the family could not cope with the symptoms of the veteran's psychological trauma. The ESWS did construct facilities to keep veterans out of asylums as it increased fundraising through the 1920s.

The public treatment on both sides of the Atlantic of veterans suffering from shell shock was both sympathetic and hostile. Those veterans who received serious physical injuries as well as having psychological trauma were viewed more sympathetically, as if they had earned the public's sympathy due to their injuries. They could not be viewed as cowards. On the other hand, the culture still embraced masculinity and bravery, and the public wondered if some veterans might not be using shell shock as an excuse for bad behavior and decision-making. These men might be less of a man or weak in the public's eyes.

Orthopedics after the War

In the wake of the war, orthopedic surgeons took their new work, especially with traction and splints, back home to their civilian practices and hospitals. The war trained over 400 doctors in the new field of orthopedics.[23] At the end of the war, French surgeons organized their own orthopedic society.[24] Prior to the war, there were only "approximately 100 self-identified orthopedic surgeons (all male) practicing in the United States."[25] These practices were only generally found in large cities and dealt mostly with children with congenital issues. Although the growth of orthopedics as a specialty in the United States also went in hand in hand with the idea of rehabilitation of soldiers (as detailed by Beth Linker), the fact was that injuries from automobile crashes or workplace injuries could receive better outcomes if an orthopedist was involved in treating wounded limbs. The American Orthopaedic Association states that "the success of orthopaedic care in the US is closely tied to the treatment of wounded soldiers in WWI (1914–1918)."[26] This is a medical specialty that would see tremendous growth in the years after the war, particularly in the United States, as the civilian need for orthopedics became clear. The growth, though, was not purely linear on either side of the Atlantic, as Roger Cooter has described in his detailed volume on orthopedics. Instead, the initiatives of the postwar period among orthopedists were important in continuing the momentum of the specialization.[27]

A great deal of the work of orthopedics overlapped with cases that would instead be given to General Surgery. One of the hopes coming out of the war was the idea that certain cases would always be routed to an orthopedist as they could have the greatest impact on a positive treatment plan. This involved cases of amputation and issues with ligaments, tendons, and joints, whether congenital or due to injury. One of the difficulties after the war was over involved finding "enough orthopedic surgeon who were well qualified to deal with the acute and chronic surgical problems as well as to supervise the detailed and often tedious special treatment" of patients after surgery.[28] It had become clear during the war that supervision of these cases was key because of the need to keep splints in particular places and tension in a pulley system specifically aligned to allow the best healing of a limb. Certain hospitals in the United States were designated for overseas (war veteran) orthopedic cases, with one of the largest being in Des Moines, Iowa.[29] Some of these hospitals had full orthopedic departments, whereas others may only have as few as three doctors assigned to orthopedic care. Doctors had begun campaigning for the need for military orthopedic hospitals as soon as the United States became engaged in the war, based on the number of cases the French and British were already seeing.[30] As the need for orthopedics grew after the war, so did the number of such surgeons. Doctors, such as H. Winnett Orr, and their principles of practice "have since had a considerable influence on civil as well as on military surgery."[31] Today, orthopedic practices can be found even in small cities. Over 25,000 orthopedists are practicing in the United States in the second decade of the twenty-first century.

Plastic Surgery after the War

The specialization of plastic surgery may have had one of the greatest surges in interest and applicability following the war. "The large number of cases of facial trauma that began to pour into hospitals in the early months of the war enabled surgeons to gain experience in plastic operations, standardize treatment and plan operative procedures with a more predictable outcome."[32] Plastic surgeons had been a small group prior to 1914, but the great need evinced by the war would give the specialty new doctors and more prestige in the 1920s and on. Although fortunately there was no war for the next two decades, plenty of civilians would need the work that these surgeons did. Car accidents or workplace accidents most closely resembled the causes of such patients from the war.

It was not too long, though, before civilians began seeing plastic surgery not simply as a type of work done on those injured but as work done to "improve" one's appearance, although there was no initial injury.

Much of the specialization in plastic surgery focused on wounds to the face. Illustrations 5.1 and 5.2 show before and after images of a patient treated with plastic surgery. The French soldier depicted had a craniofacial injury and the damage to his lower face was greatly improved

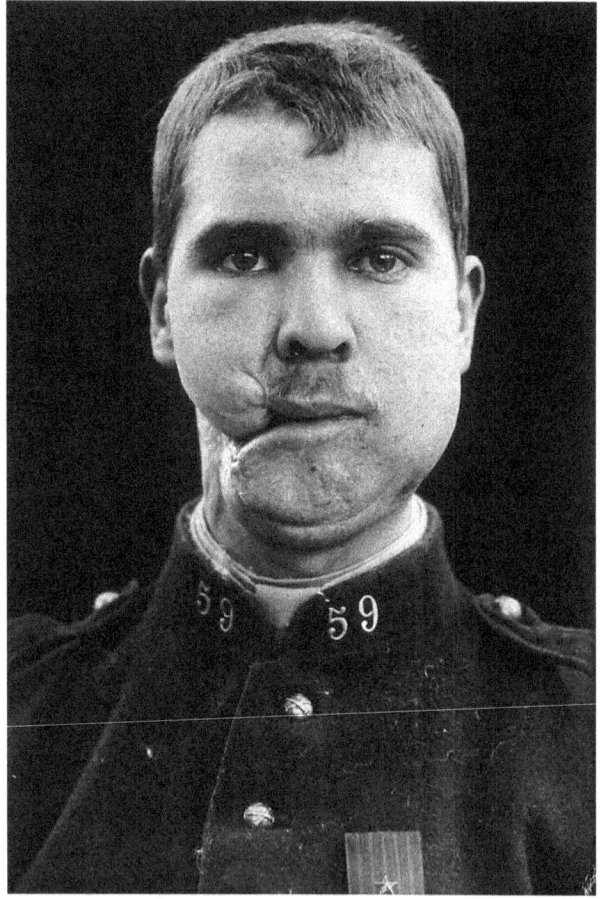

Illustration 5.1 A French soldier before having plastic surgery performed.
Photo Credit: 'Cranio-facial injury: a French soldier after incomplete plastic surgery to the lower face and neck. Photograph, 1916.'
Credit: Wellcome Collection. CC BY https://wellcomecollection.org/works/m3dvuz2y

Effects of the Medical Front in the Great War 119

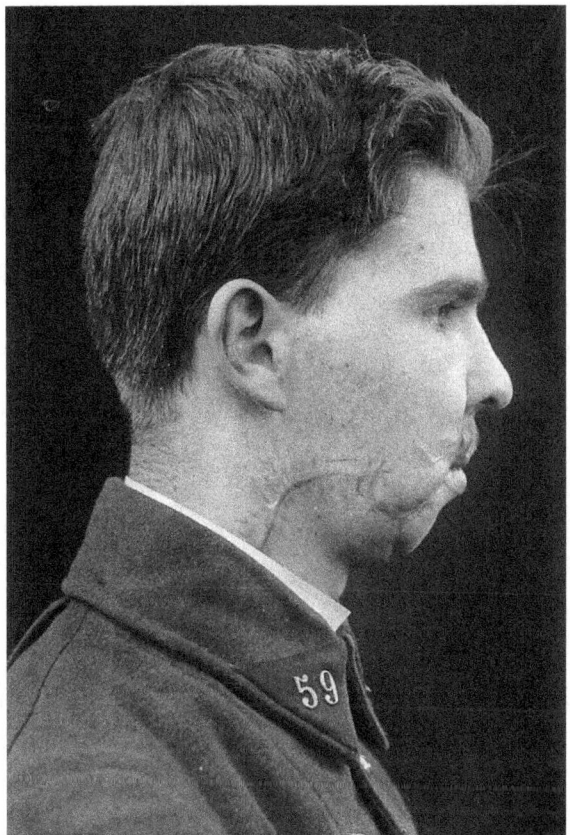

Illustration 5.2 A French soldier after having plastic surgery performed.
Photo Credit: 'Cranio-facial injury: a French soldier with scarring after plastic surgery to the lower face and neck: in profile. Photograph, 1917.'
Credit: Wellcome Collection. CC BY https://wellcomecollection.org/works/r2hts3ym

through such surgery. The expansion of beds in hospitals in Britain during the war in order to deal with these patients demonstrates the need for this type of treatment. "Elective jaw surgery was uncommon before the First World War."[33] Doctors had little opportunity to practice new techniques because there had been such a small number of potential patients, but the merit of those performed during the war was quickly appreciated. The British Dental Association began discussion of jaw and face injuries at a conference in 1916, but few could speak of exactly how bone grafts, for example, could work in these cases because of the lack of experience. Knowledge that could be promulgated was soon gained. US surgeons performed over 100 bone grafts in cases of jaw injuries in 1917–1918.[34] The work done by the plastic surgeons and dental surgeons during the war was duly compiled in medical accounts published in the 1920s and 1930s for surgeons. In 1933, a comprehensive history of how to treat fractured jaws was written.[35]

These types of works spread the word of efficacy of bone grafts and what could be accomplished through plastic surgery. During World War I, these grafts had success in 76 percent of cases. Building on this initial work in the war and the practice honed in civilian medicine in the interwar years, in World War II, surgeons had a success rate of 97 percent.[36] A similar new wave of understanding and working with cartilage grafts came in concert with the expansion of success with bone grafts. Bone grafts continued to be done on war veterans in the United States at military general hospitals in the months and years after the war. Both the value of adding mechanical stress to help strengthen the grafted bone and using different types of braces to stabilize a grafted area were expanded on and resulted in better outcomes.[37]

Aside from the growth of bone grafts as treatment, another type of work pioneered during the war also gained traction in civilian medicine in the postwar period. Surgeons found using tube pedicles could aid in facial reconstruction. This involved cutting strips of skin from areas near the face (such as the neck) while maintaining the blood supply to the strip. The strip of skin could then be reoriented to help shape and cover areas of the face. One example of their use would be to reform a nose that had been severely injured. This treatment was time consuming, and patients would not have the full results of it for up to a couple of years. However, in lieu of large tissue transplants (which weren't possible at this time), it was the only way to place new tissue with subcutaneous fat to grow over a wound.

This type of treatment was actually performed independently on both the Eastern and Western Fronts in 1917 by both a Russian surgeon and a British (originally from New Zealand) surgeon. It was promoted, however, by the British surgeon Harold Delf Gillies. Gillies "taught his techniques to many surgeons from all countries of the world."[38] After the war, Gillies was responsible for establishing plastic surgery as its own specialty in Great Britain. Gillies and one of his colleagues "in the years following the First

World War, despite the lack of after-care facilities and the difficulties of operating in unfamiliar surroundings . . . would go anywhere to advance the reputation of plastic surgery. . . . With the drop in military patient numbers this mobility was necessary to keep the specialty alive as well as provide some sort of service for the rest of the country."[39] In the few years following the war, thousands of operations for maxillo-facial injuries were performed in Great Britain. Gillies continued to perform surgeries, do hospital work, and train new plastic surgeons for the next four decades.[40]

However, it was not easy to sell the overall medical community on the benefits of plastic surgery in peacetime. "They couldn't see the need for it, and were uncomfortable with the idea of surgical operations that did not involve the curing of disease or the removal of some offending organ."[41] Many in the medical community reserved their opinion of whether such a specialization was truly needed and served a real purpose or would turn out to be used for purely cosmetic reasons that were judged as unnecessary and possibly unnatural. As Murray C. Meikle rightly points out in his history of plastic surgery, doctors failed to understand or ignored the psychological benefits of plastic surgery in these types of assessments.

Interestingly, some of the expansion of plastic surgery that occurred in the interwar period in Great Britain was also performed by a second New Zealand surgeon, Archibald McIndoe, a cousin of Gillies. McIndoe finished medical school after the Great War had ended and came to Great Britain in the 1920s to work in the plastic surgery unit that Gillies had created. McIndoe later served as the consultant plastic surgeon to the Royal Air Force in World War II, during which the continued need for facial reconstruction of soldiers was evident.[42]

The situation in Great Britain differed, though, from the way plastic surgery was approached in the United States. This could be for several factors, but it has been noted that in the early twentieth century, "American medicine was inventive and energetic in promoting new specialties... "[43] In the United States, several key surgeons saw the potential in this new specialization in the early 1920s and that there was room for it in the medical establishment. Several training programs were established by experienced surgeons under the apprenticeship model. Although most of the early work continued to be concentrated on injuries to the head and neck, expansion of the specialization to deal with other bodily injuries that needed correction would happen in the 1930s. In 1937, the American Board of Plastic Surgery formed and required that its members receive two years of general surgery and two years of plastic surgery training. And it stipulated that plastic surgery not be seen as merely a face and neck field.[44]

Other organizations had already formed in the 1920s, including ones that focused on dental and oral work in the treatment of patients of plastic surgery. By 1931, there were two national organizations in the United States devoted to plastic surgery. In comparison, Great Britain did not have its first such society until 1947.[45] The *British Journal of Plastic Surgery* soon appeared, to provide an outlet for further advancement of the field.

One component of plastic surgery work where there was not much furthering of knowledge in the immediate years after the war was with skin grafts. Skin grafts were seen as the best treatment option for burn victims at the beginning of the twentieth century, and such grafts are still done today for burn victims. However, comparatively few skin grafts were done during the Great War as few burn victims survived to need such treatment. Nevertheless, new dressings for skin grafts were developed in France during the war. However, the treatment itself was much more common in World War II, and it has become a very common treatment for burns and other types of severe wounds.[46]

War Medicine in Civilian Practice

Clearly, many new techniques had been developed during the war, as well as advanced learning about proper treatments. The Carrel-Dakin method for cleaning wounds, described in Chapter 2, proved to be widely popular after the war. One doctor described it as "one of the most important and far-reaching contributions of the war to the armamentarium of the surgeon."[47] Debridement, for everything from large scrapes to severe cuts and injuries, is one of the unsung heroes of medical treatment to emerge from the war. Even once antibiotics appeared on the scene to help combat infection, debridement and proper cleaning has still been practiced. The war demonstrated the importance of these steps for wound healing of any type.

The work on blood transfusions during the war became obviously applicable to civilian medicine. The anticoagulants experimented with during the conflict meant that blood transfusions were viable treatment options for severe blood loss in hospitals after the war. The science of making the blood usable was an outgrowth of the war, but so too was the knowledge that blood transfusions helped when patients were hemorrhaging and had already gone into shock. Before the war, doctors surmised that blood transfusions would help in such a circumstance, but the war proved the value of the practice. This was especially true for the medical scene in Europe, where transfusions (which had been direct from donor to patient) were uncommon before the war, so "by the war's end its widespread utilization for haemorrhage was a doubly dramatic change."[48] The utility of blood transfusions for civilian patients would be more fully explored after the war. Using gravity to encourage the flow of blood from a stored unit to a patient was introduced in 1935 in Middlesex Hospital in London.[49]

It must be pointed out, however, that in the case of blood transfusions, the techniques and usage seen in the war were not immediately embraced. There had been hemolytic accidents in transfusions during the war. Different blood-type testing was available, and doctors did not all use the same test. Blood-type testing may not have even been done in certain critical situations because the injuries might mean the patient would expire shortly, so doctors gambled on a blood transfusion, any transfusion, quickly. What was clear was that for the civilian population, there must be a controlled donation and typing of blood, clear and reliable storage of such, and protocols for when and how blood transfusions would be given.

As military doctors assessed circumstances as the Second World War approached, the value of blood transfusions was clear. One military doctor discussed that the military must understand that "extensive use would be made of blood transfusions" and that refrigerated blood must be available in forward positions to be given in "drip transfusion" in future conflicts.[50] Blood transfusions would become a standard in both military and civilian hospitals. The growth of blood banks, where donor blood was stored to be used in the future, first began in the late 1930s and grew in popularity and need after World War II.

Many of the various realizations and new modes of treatment from the war were still being detailed and advocated for in the decades after the end of the war. In 1939, a series of lectures was delivered at the British Postgraduate Medical School on the topic of war wounds and air raid casualties. One focused on how wounds had been treated in combat, and it detailed how different wars and their environments can affect the types of injuries soldiers had. So, in the Boer War in South Africa there was mobile warfare and a fairly sterile environment, but in the Great War, the fighting was stationary, with a high risk of infectious agents. Not only were many of the lessons that were learned in the Great War detailed for the next generation that would fight, but also it was well-defined that these lessons would be utilized by civilian medicine as well.

Controlling bleeding and preventing shock were clear aims from the war that could be applied in any number of circumstances in civilian medicine. Using morphine to keep the patient calm and help prevent shock was another. The use of the Carrel-Dakin method of irrigation was also discussed, as well as the setting of wounds according to the Winnett-Orr treatment. Dr. Orr's work on orthopedics in this war has already been detailed in this book. The importance of immobilizing fractures and making sure that patients were transported in appropriate positions were key to successful outcomes. So, that the patient with a facial or jaw injury should always be moved in the prone position is yet another lesson learned in the war.[51]

As more and more surgery patients flooded in for assessment, the importance of stabilizing them became more evident to the doctors and nurses treating them. This did not mean simply treating the symptoms of shock the patients exhibited (though some of the treatment was similar) but also, in effect, priming them for surgery. The medical staff made sure patients were given fluids and properly hydrated. They made sure that patients were properly warmed. They walked a fine line managing blood pressure and IV fluids to stave off having too much acid in the blood, which could cause irregular heart rhythms. The staff would work to get the patient in as good a condition as possible so that the patient had the best chance of surviving surgery. "There was also a developing understanding that a period of resuscitation prior to surgery was vital. This is important to optimise the patient's general condition … before proceeding to surgery—something later generations of surgeons had to relearn."[52] Likewise, once surgery was begun, it must be completed as quickly as possible. Surgeons saw during the war that those who performed more slowly had patients with poor outcomes. These types of protocols became incorporated into civilian medicine following the war.

Numerous head wounds were suffered by soldiers in the war. Neurosurgery was in its infancy, but surgeons like Harvey Cushing, who served with the Allied armies, brought a long history of dealing with brain injuries to their positions near the front. In the wake of the war in 1919, Cushing delivered an address describing a new technique and new equipment to deal with brain surgeries. A device had been created that delivered a "high frequency electric current" to blood vessels to cause coagulation and limit bleeding when doing brain surgery.[53] This dealt with one of the largest problems in doing brain surgery, which had been the massive loss of blood such surgery triggered. Cushing oversaw numerous cases in France as this device was developed and perfected. After its introduction, surgeons used it not only in brain surgeries but in other types of surgeries as well.

Other types of advances did not mean new techniques or inventions; instead, it was a change in method and approach. It was during the Great War that medicine began to fully understand concussions and the risks associated with them. "One of the achievements of the wartime experience… was the lesson that apparently minor head wounds carried potential lethal complications."[54] Patients would walk into a hospital, seem relatively fine, be assessed with having only a minor head injury, and yet go on to have major complications that might even result in death. Doctors realized how impossible it was to accurately diagnose a head injury quickly. Instead, they needed to watch the patient for many hours to begin to put together what that patient might be facing. This change in approach became a hallmark of civilian medicine and the treatment of head injuries. "It became routine in the modern era of neurosurgery to detail patients… for a period of at least 24 hours observation," so that if the patient deteriorated, it would be seen, and intervention would be quick.[55] It is common protocol for individuals today who may have received a blow to the head to be observed in such a fashion still.

Influenza

The seemingly unrelenting wave of influenza continued after the war ended, and the medical service people returned to the home front now needing to deal with this phenomenon. Although the flu obviously presented a major difficulty within the military ranks, it posed even more problems for the public at large. Within a military organization, there is an authority present that commands and people obey; not so in the public arena. Instituting quarantines and closing establishments could not be passed off as a necessary war measure. Instead, the hope to control the virus had to be balanced against public fear and hysteria.

As the initial months of the flu epidemic passed and officials realized the scope of the crisis, they also confronted the fact that they faced a shortage of doctors and, most acutely, of nurses due to the Great War. In the United States, the Red Cross worked persistently to mitigate these shortages.[56] The US Public Health Service also sought to control the response to the disease outbreak, bringing more attention to the role of public health in society. "The power of local public health agencies varied widely," and this contributed to what could be an uneven response to the crisis in different parts of the country.[57]

The true horror of this epidemic can be seen in the totals of those who perished. Over 50 million people died worldwide. Approximately 550,000 died in the United States, with millions sickened.[58] Though public health services on both sides of the Atlantic worked diligently to stem the tide of the virus, many have pointed out that their success was not consistent. Coming on the heels of so many new scientific understandings and breakthroughs regarding disease since the 1890s, this epidemic was a frustrating experience for many medical personnel about the limits of their expertise.

Medical Bureaucracy and Public Health

Any large organization develops its own bureaucracy that has its own rules, protocols, and perspectives. With hundreds of thousands of people to provide medical care for, the various medical corps were initially situated to affect the course of public health initiatives that could also translate to a wider society in the future. Another type of carryover from the medical front to the civilian world was the bureaucracy of medicine. Public health was coming into its own in the early twentieth century, benefiting as much from the rise of progressive reforms as from the scientific understandings of the germ theory of disease. If diseases were spread by bacteria and viruses, it gave much more rationale to having a robust public health program that targeted these diseases and promoted education. Public health initiatives, therefore, could have a real impact on the course of disease in the human population.

As much as people desire to find rationality in action, there was precious little of it during the war. The use of vaccinations to prevent disease had been around since Edward Jenner began his experimentation on smallpox and cowpox in the 1700s. Vaccines had been seen as risky for many decades following their introduction, but they had become more accepted as needed prevention for a number of diseases in the early twentieth century. In fact, they would ultimately be instrumental in the eradication of several diseases that had plagued humankind. As many successes as the medical field had during the war, there was still room to have done more, and this was certainly the case for vaccination usage. For example, the French and Italian armies made typhoid (fever) vaccinations mandatory for soldiers in their armies. However, the British Army refused to make it compulsory for its troops. As G. Dennis Shanks accurately points out, "The British Army regularly ordered men to charge across no-man's-land to assault heavily defended trenches, but would not order the same men to receive typhoid vaccine—a scenario that shows the degree of disconnection between medical science and military logic during World War I."[59]

Yet, the statistics that do exist on typhoid vaccination demonstrate its clear efficacy during the war. In the US Army, typhoid rates fell from 142 cases out of 1,000 soldiers in the earlier Spanish-American War of 1898 to less than 1 case out of 1,000 soldiers in World War I.[60] Similar numbers followed for the French and British armies. In that series of lectures in 1939, one commentator mentioned that steps were then being taken to make sure that the modern British Army be properly vaccinated. There were comprehensive efforts to prevent tetanus through a vaccination process. Tetanus bacteria were present in the farm fields of France, and early cases of it during the war had a high mortality rate. In the fall of 1914, the medical corps for the various armies began administering anti-tetanus serum to the wounded soldiers, perfecting the dosage over the next couple of years. The serum had been developed in the 1890s, but "it was only during WWI that the merits of tetanus antitoxin treatment… were fully recognized."[61] In 1918, the British alone estimated that 10 million doses of anti-tetanus serum had been dispatched to wounded soldiers.[62] This serum was "one of the most successful preventive inventions in wartime medicines."[63] The benefit of vaccinations was becoming accepted. Vaccinations in civilian life became increasingly commonplace and even mandated by governments and policies. The public health impact of wartime medicine would be felt for decades.

The bureaucratic nature of the medical side of the war was clearly documented in the development of the various medical corps for the Allied forces. This too made its way into civilian life through such institutions as Great Britain's Ministry of Health, which was established in 1919.[64] The general health of the civilian population became a greater focus for governments, though this would vary among countries. The modern concept of large hospitals with an equally large staff and the referring of patients to specialists grew out of this period as these physicians saw the benefit of working in such an environment both for themselves in obtaining greater knowledge and better procedures and for the outcomes for their patients.[65] The future would be different, as Roy Porter observed regarding the medical organizations constructed during the war: "Though such medical machines were dismantled after the armistice, outlooks were permanently changed."[66]

A strong bureaucratic emphasis on sanitation demonstrated a success of the growing medical bureaucracy. Delousing, quarantine, and better hygiene could prevent the spread of some diseases. The medical corps realized that the prevention of disease was more important than treating it. Part of the reason the medical specializations could flourish, especially with regard to the US Medical Corps, was due to the creation and success of the Sanitary Corps. With the exception of the influenza virus late in the war, doctors did not face the same epidemics as doctors had in previous wars.⁶⁷ It should be remembered that until the Great War, deaths from disease always had outpaced deaths from combat.⁶⁸

The Sanitary Corps in the US Army was tasked with maintaining hygiene standards. It also served to complement the work of the Medical Corps in lessening the effect of disease by such efforts as mosquito extermination, and it took on the task of food nutrition for the soldiers. The Sanitary Corps was to see to the overall health of the uniformed soldier and the "hygienic requirements of large military communities."⁶⁹

"World War 1 marked a key transition towards scientific medicine incorporating and understanding that infectious diseases are caused by microorganisms and therefore [are] susceptible to rational control and treatment measures even in the pre-antibiotic era."⁷⁰ Protocols that were developed to prevent infection during the Great War are not only useful in a pre-antibiotic era but even in a post-antibiotic era. It is always better for the patients if an infection never happens. They simply have better outcomes. The concerns about antibiotics being contraindicated with another medication the patient is taking are not an issue. The concern of the antibiotics becoming overprescribed and losing effectiveness are similarly diminished.

Another arena that medicine became involved with is the fitness for service of soldiers. Psychological testing began in the United States during World War I as a means to select those best suited for service in the draft and to rule out those who did not have the mental capabilities for service. But there would be other medical testing that would be expanded after the war. Physiological testing for those interested in aviation, for example, would come under the purview of military doctors. Certain health standards would be imposed, particularly in training camps to promote hygiene. These would be "entirely new subjects in military administration."⁷¹

The US military considered one of its major problems to be soldiers becoming unfit for duty because of venereal disease. Therefore, numerous programs focused on educating soldiers about sexually transmitted diseases and how to prevent them. Propaganda posters encouraged soldiers to avoid situations where such diseases might be present. Red-light districts near US military bases were shut down in 1917, and even bars were not allowed in close proximity to such installations. The campaign for

STD prevention continued in France, with US soldiers forbidden from visiting French brothels.

One surprising outgrowth of the medical front was the realization by many physicians that there was much to be gained by working with one another in collegial settings. This would affect the way health care was delivered after the war. It became more common, particularly in certain regions of the United States, that doctors began setting up group practices rather than establishing an individual practice. Working in situations "in which they had backup" as they had in the war became a common method for doctors for the rest of the century.[72]

The role of medicine in promoting the public health of the military during the war provided a strong inclination to see the role of medicine in promoting the public health of civil society after the war. The Great War confirmed "that health was a national concern."[73] Just as the health of the soldiers meant a stronger army, the health of a population or workforce would mean a stronger nation, a stronger economy. "Manufacturers… are recognizing as a necessity the protection of the industrial workers. The newer methods of treating injury, as developed by the war, will prove of the utmost advantage in caring for the accidents of industry."[74] Diseases that had waylaid workers could also now be more contained and controlled, resulting in better worker output. Clear attention and initiative were fundamental to this. The mortality rate of the troops in France due to disease was lower than that of the US civilian population.[75] If the medical community had the support and time to pay attention to the civilian population in the same way that they had to the military population, great strides could be made in the nation's health.

An intrinsic worth was now assigned to promoting public health that had not fully existed before 1914. As Charles Mayo noted, "The records… show a startling lack of interest on the part of our national, state and local authorities in their responsibility fully, and often even partially, to protect the people against preventable diseases and the accidents of industry. It is a poor government that does not realize that the prolonged life, health and happiness of its people are its greatest asset."[76] This would mean a continuing intertwining of medicine and government. "We of the Medical Corps came out of the medical profession; it is our desire to keep in closest touch with the profession, working together, in the future as in the past, for the common good of our country."⁷⁷ There were numerous proposals about how this could be accomplished, from including a cabinet-level position to advise the president of the United States on matters of public health, to changing government funding to tackle specific health problems in the nation.

In the years after the war, public health initiatives in the United States were filled with starts and stops. Although public health obviously includes a national agenda to promote health, it also encompasses health or medical insurance that would facilitate paying the fees that doctors charged. During the war, perhaps filled up by the heady prospects of the impact of better medical care, the American Medical Association (AMA) was in favor of compulsory health insurance. However, this enthusiasm soon failed in the aftermath of the war as physicians and others began to fear what "socialized medicine" would look like and that it was even un-American comparing it to Bolshevik programs. The AMA was clearly against compulsory health insurance after 1920 and took on a further conservative bent in the following years, decrying that provisions such as federal funds for programs for mothers and children would not work.⁷⁸ In the 1930s, however, the New Deal funneled monies into programs to promote health and the building of various institutions for the sick and infirm.

In Great Britain, a National Insurance program had been adopted before the war. Though controversial in some circles, it did provide a level of state interest in the health of the citizens. After the war, various schemes were proposed to further public health in the nation, but many of them fell victim to the financial hardships of the interwar years.

Shortcomings in Medicine

Although many successes came out of the medical front, there were also plenty of examples of medical problems that defied treatment at the time. Despite the attention and work of doctors and nurses, gas gangrene continued to prove fatal much of the time. Sometimes these fatalities had to do with the location of the infection and the inability to amputate in that region, but even with excising the tissue, if any of the infection remained, it was unlikely to be remedied. There was no medical solution forthcoming in the 1910s. Although gas gangrene is not as common among the general population as it was among the Great War soldiers, it still does occur. Penicillin became the first drug to effectively treat it. In the twenty-first century, the same early steps of debridement, cleaning, and possible amputation are still part of the potential treatment protocol. Penicillin and other antibiotics are also used to combat the infection. These drugs have significantly improved the outcomes for patients suffering from this particular infection.

Another arena that defied improvement in the 1910s and 1920s was pain management. In the twenty-first century, pain management is still an area that needs further investigation. Today, patients do not always feel that their pain is being taken seriously by doctors, and doctors often worry about prescribing narcotics that could be addictive. Pain has proven to be a major obstacle with regards to rehabilitation after an injury. With the large numbers of amputations during the war, tens of thousands of patients had to deal with post-amputation pain. This generally appeared in the form of either stump pain or phantom limb pain.[79]

Surgeons during the war understood that post-amputation pain was a difficulty for many soldiers, and accounts and reports after the war among surgeons reiterated the idea espoused in January 1918 that a number of these patients were "healed not cured."[80] Pain did not only occur immediately following an amputation but could also reoccur even years later after a period of a patient having no pain. It was frustrating to doctors not to be able to help their patients with this complication; however, these surgeons generally discussed this only among themselves and their medical societies. They did not reach across medical specializations to see if anything new was happening with other conditions that might have a bearing on the post-amputation pain problem. The narcotics available to patients suffering from such issues changed through the twentieth century, but the problem has evaded easy solutions.

There were also gaps in treatment—treatment that should have happened for the wounded but didn't. Perhaps it was because of the dearth of doctors in a particular location, an overtaxed nursing staff caring for too many, the total number of patients, or the disinterest of the medical staff regarding a certain type of wound or illness. The triage process was efficient, but humans do make mistakes, and patients could be labeled as more injured or close to expiration than they truly were. Because of the crowding in hospitals at times, patients might only rarely see a doctor, and the doctor's assessment was only cursory and not complete. Accounts such as that of US soldier John Lewis Barkley, who was hospitalized with various illnesses and was rarely seen by a doctor, are testament to this happening. Whatever the reason for these types of gaps, this would need to be improved on in future wars. Standard protocols that medical staff followed would help with this phenomenon.

Though many advances did occur with the specialization of plastic surgery, the surgeons could not fix all injuries. This could happen based on the patient's own decision to no longer proceed with more surgery or it could be the doctors' opinion that, based on tissue loss, there was nothing further they could do. And patients did regularly decide to discontinue treatment. There were some patients that had multiple surgeries from 1917 into the 1920s in the cases of severe facial and jaw injuries. Once functionality was restored, the continuous rounds of surgery for what might be described as "cosmetic" reasons could be hard for a patient to take. In these cases after the war, masks or external prosthetics became the favored way to mitigate obvious facial injuries. Again, this spoke to the psychological need of the patient to appear "normal" or "whole," but it also reflected the true burden surgery and recovery could have on the patient.

Such masks were developed on both sides of the Atlantic. Chapter 2 mentions the work of Anna Coleman Ladd, who was a US sculptor who opened a studio in Paris that made masks for soldiers during the war. British hospitals also employed artists and sculptors to design and paint these lifelike masks, which were often done in copper and silver after obtaining plaster-of-Paris molds of the patient's face. Although these masks helped, they did present their own challenges. They had to be held in place in some way, usually by being attached to eyeglasses. They could slip and fall off unexpectedly. During the war, large bone grafts or large tissue grafts were thought impossible, but in the years after the war, as these types of grafts became more practicable, some of these types of injuries that masks were created for could then be fixed.[81]

Although there were several advances made in the field of neurosurgery during the war, some techniques still proved problematic. When surgeons needed to access the brain to remove shell fragments or bullets, they did so by using a trephine. Essentially, this device cut a hole in the patient's skull that was removed to provide access to the tissue underneath. The problem with this was that if the hole was not large enough, the surgeon then chiseled out a larger one, which meant that the original bone could never be put back over the hole and healed. Though this had been a known problem for years, other techniques were considered too complicated or took too long to perform, so this problematic method was still used in the Great War.

Conclusion

It is important to clarify that the war by itself did not stimulate medical innovation out of a vacuum. Rather, it provided the opportunity to implement new procedures and perfect others on a time line that would not have been possible in peacetime.[82] It is the acceleration of these practices that the war provided for, as well as why it was instrumental in adding to the medical understanding of the human body. "Dynamic and functionalized understandings of the body" allowed advances in preventative and therapeutic care long after the war ended.[83] For example, in the 1920s, the applicability of debridement for civilian surgery was considered slight. However, for decades now debridement is one of the first tasks doctors perform on an open wound, realizing that it helps minimize infection. Lessons learned on how the stomach and intestines healed in World War I allowed progress to be made in the years after the war on gastrointestinal disorders. Doctors knew these wartime practices would be useful, but they may not have had the foresight to truly gauge how they would be utilized. After the war, the surgeon general of the United States opined that "these surgical principles, the most important which came out of the war, need not be lost to civil surgery, but may indeed be of the highest practical value in industrial and railway surgery."[84]

Doctors also redeveloped their profession as a result of the war. An article near the end of 1918 in the *American Medical Association Journal* argued that the profession should band together after the war "to secure the rights and recognition" due it and "to have that real influence necessary for the best interest of the public health in the new order of things."[85] Some scholars have gone so far as to even note that in a way civilian medicine is militarized because of war. Many people realized the benefits that would accrue from the war's medical front lines.

A major factor in the proliferation of this medical knowledge after the war was one of the tenets of progressivism especially as practiced in the United States. Progressives emerged in the last years of the 1800s as reformers hoping to correct the excesses of industrialization that had occurred by creating more protections for people. Workplace safety rules, getting rid of corrupt politicians, and the breaking up of business monopolies were all parts of the progressive message. It was a movement with a broad umbrella, however, and progressives also sought to "clean up" other arenas that had seen abuse and corruption. The notion of a doctor being a snake-oil salesman was one such target. The standardization of curriculum in medical schools, the practice of creating associations of those in the same career fields (orthopedics, etc.), and the belief in expertise being deferred to all gave an avenue to these practitioners' spread of relevant treatment protocols across several nations. In the United States, "the changes constituted a veritable revolution in medical education." The number of medical schools stood at 162 in 1902 but numbered only 76 by 1930.[86] The faith in experts that was seen by progressives gave further substance and credibility to the medical profession during the war years and afterward. The development of specialized training programs at certain hospitals meant that specialists would continue to be trained by mentors for civilian practice.

The military medical corps of the United States and Great Britain did achieve advancements in other ways during the war. The scientific medicine achieved during the conflict later became an integral part of the modern state's efforts to ensure the health and continued productivity of its working population.[87] The lessons learned in World War I would be utilized again in World War II when it came to the wide-scale warfare with mass numbers of troops and severe injuries again. World War I proved the "value of cooperation with the medical profession in relation to any future mobilization for war."[88] Though separated by 20 years, the Great War was a staging ground for many of the modes of operation that would be seen in World War II. In fact in 1939 as the Second World War approached, one military doctor mentioned how so much of modern warfare's medical side could be reapplied from World War I, but one of the questions was what would be the true effect of air power in the next conflict, as that was not seen on a large scale in the Great War, and therefore there was no precedent for it in the medical community.[89]

One of the key lessons of the Great War was in the role organization played. In the 1920s, the goal became for the US Medical Department "not merely to institute educational training of enlisted men as clinical clerks and surgical dressers in the hospital, but also to train specialists among… commissioned personnel in all the important branches of scientific medicine.… The older, archaic army of the small, scattered and isolated posts disappeared with the European war."[90] Improving the education for those in military medicine would certainly help when another war occurred. One plan proposed, but not enacted, would see 10 percent of the graduates of medical schools go into military medicine training with rotations in public health, sanitation, field service, and hospital work. The hope would be that this would enable the military to have a trained reserve medical corps in place, familiar with the norms of the military service and ready for the next conflict.[91]

Yet, for the clear strides that medicine made in the war for the betterment of its patients, not everyone fully understood the role it played. Charles Mayo, one of the founders of the Mayo Clinic and one of the respected and well-known medical practitioners of the 1910s and 1920s, wrote about what he saw as a lack of credit being given to the US Medical Corps. He explained how integral the medical services were to the strength of the army. Giving accolades to the Medical Corps, he stated, "The real strategy of this war has been the control of disease by the men in the medical corps who reduced the loss of man power by preventive medicine, and by treatment controlled diseases that in the past were determining factors in the defeat of armies."[92] Despite the positives to emerge from the work of the Medical Corps, Mayo went on to lambast how the government and military in the United States treated these people:

> It is deplorable, and a blot in the history of this war, that the work of the American medical profession was not recognized by the general staff

and that rank, according to responsibility, was withheld in most instances until the war was nearly or quite over. A comparatively long period of service in Washington enables me positively to state that the same unfairness toward the medical army officer still exists.[93]

What infuriated Mayo was that the US military provided plenty of training for their own officers either initially through a service academy (West Point or Annapolis) or through continued training once commissioned. Yet, there was no aid or training forthcoming (or even officer rank) for those in the Medical Corps. Instead, the military and the government benefited from the work and study of these doctors and nurses who had often incurred their own expenses for their own training (Mayo cites dollar figures in the thousands), and there was so little recognition or reward for this. Mayo hoped that reorganization by the time of the next war (he did believe there would be another) would give civilian medical professionals their due when they served in the military. Mayo was clear to point out how important being given "rank" in the military was for doctors because that was the established hierarchy and system of authority there and was the only thing that would truly command respect. The way the United States approached this was different from the way the British did, as the British did give higher military ranks to their physicians.

The medical front of World War I is an important one because it provided so many opportunities for improvement of medical treatment that would benefit innumerable future patients. It served as an unprecedented time to experiment in treatment at a time when medical science had made enough progress as to be able to have real success with surgeries and treatments that even 20 or 30 years earlier would never have been thought possible. Wide-ranging medical advancements continued to come at a fast pace in the twentieth century as pharmaceuticals became a major component of treatment plans. Doctors could turn to a variety of new medicines to aid with chronic conditions and to help speed healing or to minimize side effects of treatment. The expansion of technologies from the X-ray to the CT scan or MRI revolutionized understanding injuries and illness and pinpointing treatment in newfound ways just as much as when doctors began using X-rays to track bullets. Much of the foundations, though, of "modern medicine" and the clear rise of specializations can be seen in the practice of medicine in World War I. As orthopedist H. Winnett Orr stated in the introduction of his book based on his wartime experience, "The war gave us a marvelous opportunity to learn something about accident—injury-reconstruction-orthopedic surgery. The war hospitals were an enormous laboratory for surgery. Many experiments were tried. The careful students learned much.... Improvements in the civilian practice of surgery can be brought about by a more general recognition of the lessons of the war."[94]

The doctors from World War I dispersed after the conflict, but they took with them their new knowledge and the commitment to continue to spread it and build on it in the coming decades. The exposure of nurses and stretcher-bearers to the medical front similarly enhanced the treatment of those patients who saw these people later in their medical careers. The introduction of so many, including the soldiers, to what "modern medicine" could do permanently changed views and expectations of what should be done for soldiers in subsequent wars. Medical science has advanced tremendously because of their work both during and after the war.

Conclusion

The medical front during World War I embodied both everything that was positive about humanity and everything that was disconcerting. It was filled with people who wanted to help one another, care for one another, and make an awful situation better. It was also filled with the victims of man's inhumanity to man. The types of wounds soldiers (mostly) and civilians (often) suffered in this conflict defied one's imagination of suffering. And though there were breaks in the action, the slew of patients simply kept coming. The medicine that was practiced was a direct outgrowth of the scientific knowledge that had recently been obtained and the new value of organization and efficiency that were hallmarks of the early twentieth century.

Medical personnel had to get creative to treat the wounded. And they took pride in bringing patients back from the brink of death. They documented their successes and their failures, often searching for that precise combination of skill and treatment that would render their patients whole. The nurses, stretcher-bearers, and doctors all had to find their own ways to cope through their months and years of service.

It is in part their remarkable skill at returning soldiers to the front lines that made the war's continuation possible. If these soldiers had been unable to recover from these wounds or succumbed to disease at the same level that soldiers had in past conflicts, the nations at war would have literally run out of troops at some point. The work of the various medical corps was instrumental to this industrialized warfare. The medical practitioners realized this and often lamented they were patching people up only to throw them into the maelstrom once again.

Because of the variety of wounds that were seen, new techniques were developed in a number of medical specializations. Doctors were quick to see the benefit of these techniques and hoped to expand their usage after the war, to continue treatment of the war veterans and to help others in civilian society when confronted with similar injuries.

The type of logistical organization required to put medical personnel where needed in the numbers needed took some time to put in place. The British military found the beginning of the war difficult to handle regarding medical needs. The United States benefited from having seen these fits and starts and chose to err on the higher side of the spectrum for what would be needed. These bureaucratic efforts, though, did provide a great deal of information for these same governments as they approached World War II. This is especially true for the United States, which would enter World War II in 1941 with the logistical nightmare of fighting on two fronts from the very beginning. Medical corps knew better how to handle such things as transport from an active front to an area in the rear for treatment, and how to stabilize the wounded. They had learned how to expedite cases when flooded with large numbers of patients. They understood better how to treat shock. They had a much clearer sense of how many patients they would likely be treating, how many beds would be needed. Even things such as how to better stock supplies helped managers as they prepared for another war. The governments and medical corps also had a better idea of what aftercare would look like for the thousands and thousands of wounded who would survive combat. And they had acquired the knowledge that treatment for veterans would not be only physical but also psychological.

Just as expectations had changed for what the government and military would need to do medically in another war, so too had the expectations changed for those who would be fighting another war. The soldiers obviously directly benefited from the medical care offered during the war. But the impact of that care went past the immediate injury that needed treatment. The soldiers themselves came to expect decent medical care from their militaries in the Allied countries after the war. This was nonnegotiable. It became bound up in the social contract between soldier and government. They had a general confidence that everything that could be done for them would be done for them. Families of soldiers, as well, came to expect a minimum level of medical care would be provided to these men. The fact that the military itself publicly discussed the aftercare and rehabilitation programs it developed gave further credence to the idea that this care was a soldier's due for risking their life for their country. In the year after the war ended, Charles Mayo, one of the founders of the Mayo Clinic, wrote, "We must also recall the improved morale incident to the modern treatment and quick healing of wounds, whereby men were repeatedly returned to the ranks, instead of remaining permanently disabled and becoming a charge of their country, as in former warfare."[1]

The "war to end all wars," as the Great War was called, visited a new industrialized vision of warfare on the world that meant that war would forever lose its romanticism. It was no longer something to be celebrated or anxiously anticipated by young men. The soldiers fought in relentlessly bad conditions for four years and quickly disabused themselves of the ideas of glory they had been fed as children. Relief, as much as any emotion, accompanied the end of the war for all sides.

In the years following the war, this disillusionment coalesced into a deep cynicism regarding the future. In *Mud, Blood and Poppycock*, Gordon Corrigan (2003) described how "hangovers of the war ... are still with us."[2] These hangovers were most acute in the 1920s. There was no reason to believe that the future would be better than the past. There was no reason to have faith in humanity and charity. The word *progress* could no longer be placed on a pedestal to be celebrated by a technologically driven society. Many people saw the world not as a promising place, but, as the poet T. S. Eliot labeled it, "The Waste Land."

War itself and the human endeavor in war seemed monumentally futile. And more than futile, it seemed a blasphemy. In the 1920s, many of the countries that had engaged in the Great War sought to distance themselves from it. This distancing at times took on fanciful notions such as the Kellogg-Briand Pact of 1928, which outlawed war as an instrument of foreign policy. Though an ultimately ineffective document, it had over 60 countries become signatories. Going into another war became the last thing a country such as Great Britain or the United States wanted to face through the 1920s and most of the 1930s. There was no public will for it, and it would spell the end of a political career to suggest it. The United States retreated into isolationism shortly after the end of the Great War, refraining from even approving the Treaty of Versailles. This isolationist bent stayed in place through the next decade and became more legally hardened in the 1930s with a series of Neutrality Acts deliberately built on the events that propelled the United States into the Great War. The United States would not sell items to a country at war, which would mean that US ships would not be in the waters of a war zone and subject to potential targeting by submarines, as had been the case in the 1910s. Such precautions, of course, did not work in the end, and the next world war would have the United States drawn into it as well.

The British too shied away from the prospect of war, scarred by the experience of World War I. In the 1930s, after Adolph Hitler came to power in Germany and began his rearmament of the German military, Britain professed it could work with Hitler. These violations of the Treaty of Versailles, which forbid such rearmament, need not require the British to go into another war to curb Hitler's actions. As much as the French feared a rearmed Germany, they also had no thirst for war. If there were ever a point at which Hitler's plans could have been derailed, it would have been in the mid-1930s, but the specter of the Great War, the continued "hangover" of it, had not dissipated, and there was no desire to engage in another conflict.

The myths of the First World War have persisted through the ensuing decades. Society has looked back at it as pointless, as a waste of men. As Dan Todman (2005) writes, "There existed in British popular culture a unified mythology of the First World War that depicted it in terms of mud, horror, stupidity, and futility."[3] As Todman also points out, this is by no means the whole story of the war. On the individual level, this was a very personal war with small victories hard fought for and individuals coming to terms with the limits of their endurance. Just as it is difficult for a person today to keep in mind the entire scope of the war, so too was it for each soldier or nurse during the war. Their focus had to be on what was in front of them and what they had to do.

What the fight on the medical front offers is a small redemption of the perceived futility of this war. It offers those who study it and learn from it an opportunity to be reconnected with the humanity of the caregivers and the soldiers at the very time that such human qualities were ignored in the juggernaut of the war. It also can serve as an example of how technology and societal trends can impact one another. It can personalize a conflict that is often painted with broad strokes.

The Allied armies won a victory in the war, but also among the victors were "doctors who had specialized in wartime surgery, shellshock or heart medicine [who] returned to civilian life with a passionate vision of a better medical future."[4] The war affected the professionalization of medical careers in nursing too, though nursing remained a gendered enterprise for the next decades. All those involved in medical care became greatly affected by what they witnessed during the war. Medicine itself substantially changed its protocols and treatment patterns because of medical practices during the war. Some of these changes were novel and exciting, such as the inroads made in plastic surgery, but even seemingly more mundane realizations such as how to approach people with concussions would substantially change the outcomes of numerous cases after the war. The patients who received new treatments and who pulled through were also victors over the specter of injury and death. And at the same time, it was not lost on the practitioners or the patients that these successes of industrialized military medicine allowed the industrialized military warfare to continue. The First World War dispelled much of the innocence of Western society, but it did offer a significant advancement to medicine and for patients in the future.

Notes

Introduction

1. Mark Harrison, "The Medicalization of War—The Militarization of Medicine," *Journal of the Social History of Medicine* 9 (August 1996), 267–68; and Mark Harrison, *The Medical War: British Military Medicine in the First World War* (Oxford: Oxford University Press, 2010).
2. Roy Porter, *The Greatest Benefit to Mankind: A Medical History of Humanity* (New York: W.W. Norton, 1997), 642.
3. Charles Lynch, Frank W. Weed, and Loy McAfee, eds., *The Medical Department of the United States Army in the World War, Surgery*, vol. 11, pt. 1 (Washington, DC: Government Printing Office, 1927), 121.

1 The Great War

1. Sean McMeekin, *July 1914: Countdown to War* (New York: Basic Books, 2013), 6–10.
2. Michael Clodfelter, *Warfare and Armed Conflicts: A Statistical Reference to Casualty and Other Figures, 1618–1991* (Jefferson: McFarland, 1992), 781–88.
3. Clodfelter, *Warfare*, 781.
4. John C. Burnham, *Health Care in America: A History* (Baltimore: Johns Hopkins University Press, 2015), 149–51, 206–7.
5. Burnham, *Health Care*, 176–77.

6. Theo Emery, *Hellfire Boys: The Birth of the U.S. Chemical Warfare Service and the Race for the World's Deadliest Weapons* (New York: Little, Brown and Company, 2017), 16–17.
7. Hew Strachan, *The First World War* (New York: Viking, 2003), 163.
8. Will R. Bird, *And We Go On: A Memoir of the Great War* (Montreal: McGill-Queen's University Press, 2014), 19–20.
9. Leo van Bergen, *Before My Helpless Sight: Suffering, Dying and Military Medicine on the Western Front, 1914–1918* (Surrey: Ashgate, 2009), 130.
10. Van Bergen, *Before My Helpless Sight*, 132.
11. Van Bergen, *Before My Helpless Sight*, 16–17.
12. Charles H. Mayo, *Educational Possibilities of the National Medical Museum in the Standardization of Medical Training* (Chicago: The American Medical Association, 1919), 3.

2 Allied Medical Innovations

1. "World to Benefit by War Medicine," *New York Times*, December 29, 1918, sec. 3, 4: 1.
2. Roger Cooter and Steve Sturdy, "Of War, Medicine and Modernity: Introduction," in *War, Medicine and Modernity*, eds. Roger Cooter, Mark Harrison, and Steve Sturdy (Thrupp: Sutton Publishing, 1998), 7.
3. Cooter, Harrison, and Sturdy, "War, Medicine and Modernity," 12.
4. Richard V. N. Ginn, *The History of the U.S. Army Medical Service Corps* (Washington, DC: Office of the Surgeon General and Center of Military History United States Army, 1997), 42.
5. A.M. Fauntleroy, *Report on the Medico-Military Aspects of the European War* (Washington, DC: Government Printing Office, 1916), 46.
6. "Results Obtained at an Advanced Surgical Unit," *American Medical Association Journal* 70 (January 12, 1918), 111.
7. Charles Lynch, Frank W. Weed, and Loy McAfee, eds., *The Medical Department of the United States Army in the World War, Surgery*, vol. 11, pt. 1 (Washington, DC: Government Printing Office, 1927), 297.
8. Fauntleroy, 31.
9. Richard A. Gabriel and Karen S. Metz, *A History of Military Medicine: Volume II, From the Renaissance through Modern Times* (New York: Greenwood Press, 1992), 248.
10. Fauntleroy, *Report on the Medico-Military Aspects*, 42.
11. Ginn, *The History of the U.S. Army Medical*, 45.
12. Fauntleroy, *Report on the Medico-Military Aspects*, 33.
13. T.H. Goodwin, *Notes for Army Medical Officers* (Philadelphia: Lea & Febiger, 1917), 33.

14. Gabriel and Metz, *A History of Military Medicine*, 248.
15. Goodwin, *Notes for Army*, 33.
16. A. Don, "Dressings Used in a Casualty Clearing Station," *British Medical Journal* 1 (May 6, 1916), 648–49.
17. MacPherson, W.G. (ed.), *History of the Great War Based on Official Documents. Medical Services. General History*, Vol. 3 (London: HMSO, 1924), 170–71, as quoted in Thomas Scotland and Steven Heys, eds., *War Surgery 1914–18* (Solihull: Helion & Company, 2013), 78–79.
18. E. Ann Robertson, "Anaesthesia, Shock and Resuscitation," in *War Surgery*, 97–98.
19. J. Trueta, *Treatment of War Wounds and Fractures with Special Reference to the Closed Method as Used in the War in Spain* (London: Hamish Hamilton Medical Books, 1939), 8.
20. Matthew Naythons, *Faces of Mercy: A Photographic History of Medicine at War* (New York: Random House, 1993), 118–20.
21. Lynch, *The Medical Department of the United States Army*, vol. 11, pt.1, xxxi–xxxii.
22. "German, French, and British Bullets," *British Medical Journal* 1 (December 5, 1914), 990–91; "German and British Bullets," *British Medical Journal* 1 (December 12, 1914), 1041.
23. W. Earle Drennen, "Experiences in Military Surgery," *American Medical Association Journal* 65 (July 24, 1915), 296.
24. Drennen, "Experiences," 297.
25. F.N.L. Poynter, ed., *Medicine and Surgery in the Great War, 1914–1918* (London: The Wellcome Institute of the History of Medicine at War, 1968), 6–7.
26. G. Lenthal Cheatle, "Antiseptics in War," *British Medical Journal* 2 (December 12, 1914), 1006.
27. Perrin Selcer, "Standardizing Wounds: Alexis Carrel and the Scientific Management of Life in the First World War," *British Journal for the History of Science* 41 (March 2008), 86.
28. Selcer, "Standardizing Wounds," 74.
29. Poynter, *Medicine and Surgery*, 7.
30. Selcer, "Standardizing Wounds," 91.
31. "Treatment of Projectile Wounds by Excision of Damaged Tissues," *American Medical Association Journal* 65 (July 31, 1915), 463.
32. "Primary Suture of Wounds," *American Medical Association Journal* 65 (July 17, 1915), 288; "Early Suture of War Wounds," *American Medical Association Journal* 70 (January 5, 1918), 61; "Treatment of War Wounds," *American Medical Association Journal* 70 (January 19, 1918), 195 are a few examples.
33. Gabriel and Metz, *A History of Military Medicine*, 242–43.
34. Naythons, *Faces of Mercy*, 122.

35. "Results Obtained at an Advanced Surgical Unit," *American Medical Association Journal* 70 (January 12, 1918), 111.
36. Gabriel and Metz, *A History of Military Medicine*, 240.
37. Goodwin, *Notes for Army*, 52.
38. Gabriel and Metz, *A History of Military Medicine*, 240.
39. Gabriel and Metz, *A History of Military Medicine*, 241.
40. Lynch, *The Medical Department of the United States Army*, vol.11, pt. 1, 689–95.
41. "War Orthopaedics," *British Medical Journal* 2 (October 26, 1915), 575.
42. Robert Jones, "An Address on the Orthopaedic Outlook in Military Surgery," *British Medical Journal* 1 (January 12, 1918), 41.
43. "Wounds of Arteries in the Legs," *American Medical Association Journal* 70 (January 12, 1918), 133.
44. Poynter, *Medicine and Surgery*, 7.
45. H. Winnett Orr, *An Orthopedic Surgeon's Story of the Great War* (Lincoln: n.p., 1921), 28–29.
46. Joseph A. Blake, *Gun-Shot Fractures of the Extremities* (Paris: Masson et Cie, Éditeurs, 1918), 33.
47. Blake, *Gun-Shot Fractures*, 33.
48. Blake, *Gun-Shot Fractures*, 99.
49. Hey Groves, "An Address on Some of the Problems Related to the Treatment of Gunshot Fractures," *British Medical Journal* 1 (July 15, 1916), 65.
50. Schneider, William H. "Blood Transfusion in Peace and War, 1900–1918," *Social History of Medicine* 10 (April 1997), 115.
51. F. Boulton and D.J. Roberts, "Blood Transfusion at the Time of the First World War—Practice and Promise at the Birth of Transfusion Medicine," *Transfusion Medicine* 24 (December 2014), 329–30.
52. Gabriel and Metz, *A History of Military Medicine*, 242.
53. Lynch, *The Medical Department of the United States Army*, vol. 11, pt. 1, 151.
54. Boulton and Roberts, "Blood Transfusion at the Time of the First World War," 330.
55. Schneider, "Blood Transfusion in Peace and War," 121–24.
56. Boulton and Roberts, "Blood Transfusion at the Time of the First World War," 330.
57. "Blood Transfusion in War Surgery," *American Medical Association Journal* 70 (January 5, 1918), 59–60.
58. Boulton and Roberts, "Blood Transfusion at the Time of the First World War," 331.
59. Schneider, "Blood Transfusion in Peace and War," 119.
60. Schneider, "Blood Transfusion in Peace and War," 118.
61. Alexi Assmus, "Early History of X Rays," *Beam Line* 25(2) (1995), 24.
62. Roy Porter, *The Greatest Benefit to Mankind: A Medical History of Humanity* (New York: W.W. Norton, 1997), 605–6.

63. René Van Tiggelen, *Radiology in a Trench Coat: Military Radiology on the Western Front during the Great War* (Brussels: n.p., 2013), 50.
64. Van Tiggelen, *Radiology in a Trench Coat*, 51.
65. Poynter, *Medicine and Surgery*, 8.
66. Van Tiggelen, *Radiology in a Trench Coat*, 170.
67. Van Tiggelen, *Radiology in a Trench Coat*, 171.
68. *United States Army X-Ray Manual* (New York: Paul B. Hoeber, 1919), 196.
69. *United States Army X-Ray Manual*, 305.
70. *United States Army X-Ray Manual*, 306.
71. Alexander MacDonald, "X-Rays During the Great War," in *War Surgery*, 141–43.
72. MacDonald, "X-Rays," 145.
73. Shelley Emling, *Marie Curie and Her Daughters: The Private Lives of Science's First Family* (New York: Palgrave Macmillan, 2012), 26–27.
74. "Sculptor's Art in Surgery," *New York Times*, July 3, 1918, 12: 7.
75. John D. Holmes, "Development of Plastic Surgery," in *War Surgery*, 259–60.
76. Gabriel and Metz, *A History of Military Medicine*, 242.
77. Gabriel and Metz, *A History of Military Medicine*, 242.
78. Naythons, *Faces of Mercy*, 188–90.
79. *The Medical Department of the United States Army*, vol. 11, pt. 2, 400–2, 408.
80. Lynch, *The Medical Department of the United States Army*, 397.
81. A.R. Fisher, "Chloramine in the Treatment of Wounds of the Mouth and Jaws," *British Medical Journal* 1(2872) (January 15, 1916), 87–88.
82. Lynch, *The Medical Department of the United States Army*, vol. 11, pt. 2, 458.
83. Lynch, *The Medical Department of the United States Army*, 460.
84. Lynch, *The Medical Department of the United States Army*, 464.
85. *United States Army X-Ray Manual*, 476.
86. Lynch, *The Medical Department of the United States Army*, vol. 11, pt. 2, 453–54.
87. Frederick A. Pottle, *Stretchers: The Story of a Hospital Unit on the Western Front* (New Haven: Yale University Press, 1929), 134.
88. V.H. Kazanjian, "Treatment of Maxillary Fractures," *British Medical Journal* 1 (February 19, 1916), 266–67.
89. Lynch, *The Medical Department of the United States Army*, vol. 11, pt. 1, 105.
90. Ralph H. Major, *A History of Medicine*, Vol. 2 (Springfield: Charles C. Thomas Publishers, 1954), 984.

3 Medical Personnel

1. Marjorie Barron Norris, ed., *Medicine and Duty: The World War I Memoir of Captain Harold W. McGill, Medical Officer, 31st Battalion C.E.F.* (Calgary: University of Calgary Press, 2007), 23.

2. Rae S. Dorsett, *History of Base Hospital No. 85, United States Army* (United States: n.p., 1919), 2.
3. Clarence Benjamin Francisco, *Wartime Diary of Clarence Benjamin Francisco, M.D.* (n.p.), n.p. Clendening History of Medicine Museum, Kansas City, Missouri.
4. Arthur Anderson Martin, *A Surgeon in Khaki* (London: Edward Arnold, 1915), 31.
5. Robert V. Dolbey, *A Regimental Surgeon in War and Peace* (London: John Murray, 1917), 52–54.
6. Dolbey, *A Regimental Surgeon*, 56.
7. Dolbey, *A Regimental Surgeon*, 93.
8. Martin, *A Surgeon in Khaki*, 71.
9. Martin, *A Surgeon in Khaki*, 74–75.
10. Norris, *Medicine and Duty*, 103–5.
11. Norris, *Medicine and Duty*, 106.
12. Norris, *Medicine and Duty*, 93–100.
13. Norris, *Medicine and Duty*, 313.
14. Ellen N. LaMotte, *The Backwash of War: The Human Wreckage of the Battlefield as Witnessed by an American Hospital Nurse* (New York: G.P. Putnam's Sons, 1916), 59–60.
15. Dolbey, *A Regimental Surgeon*, 58.
16. Dolbey, *A Regimental Surgeon*, 58.
17. Ian R. Whitehead, *Doctors in War* (Barnsley: Pen & Sword Military, 1998), 33–36.
18. Whitehead, *Doctors in War*, 38–42.
19. Whitehead, *Doctors in War*, 49.
20. William L. Hanson, *World War I: I Was There* (Gerald: The Patrice Press, 1982), 5.
21. Christine Hallett, *Veiled Warriors: Allied Nurses of the First World War* (New York: Oxford University Press, 2014), 37.
22. Whitehead, *Doctors in War*, 107.
23. Whitehead, *Doctors in War*, 111–12.
24. Whitehead, *Doctors in War*, 117.
25. Hanson, *World War I*, 7–8.
26. Hanson, *World War I*, 81.
27. H. Winnett Orr, *An Orthopedic Surgeon's Story of the Great War* (Lincoln: n.p., 1921), 12.
28. Orr, *An Orthopedic Surgeon's Story*, 13.
29. Orr, *An Orthopedic Surgeon's Story*, 20–23.
30. *Medical Training Camp, Fort Oglethorpe, Georgia, August 1917* (n.p., 1917), 42.
31. *Medical Training Camp, Fort Oglethorpe, Georgia*, 42.

32. "April 15, 1918 Memorandum," (n.p.), Clendening History of Medicine Museum, Kansas City, Missouri.
33. "April 15, 1918 Memorandum."
34. "April 15, 1918 Memorandum."
35. "April 15, 1918 Memorandum."
36. "April 15, 1918 Memorandum."
37. "April 15, 1918 Memorandum."
38. J. Garry Clifford, ed., *The World War I Memoirs of Robert P. Patterson: A Captain in the Great War* (Knoxville: The University of Tennessee Press, 2012), 85.
39. Charles H. Horton, *Stretcher Bearer! Fighting for Life in the Trenches*, ed. Dale Le Vack (Oxford: Lion Books, 2013), 39.
40. Horton, *Stretcher Bearer!*, 57.
41. Harry Stinton, *Harry's War: Experiences in the "Suicide Club" in World War One* (London: Brassey's, 2002), 153.
42. "April 15, 1918 Memorandum."
43. Sarah Sand Stevenson, *Lamp for a Soldier: The Caring Story of a Nurse in World War I* (Bismarck: North Dakota State Nurses' Association, 1976), 38.
44. Charles Richet, transl. Helen de Vere Beauclerk, *War Nursing: What Every Woman Should Know, Red Cross Lectures* (New York: Robert M. McBride & Co., 1918), vii.
45. Stevenson, *Lamp for a Soldier*, 42–45.
46. Ruth Cowen, ed., *A Nurse at the Front: The Great War Diaries of Sister Edith Appleton* (London: Simon & Schuster, 2012), 23.
47. Cowen, *A Nurse at the Front*, 26.
48. Anne Powell, *Women in the War Zone: Hospital Service in the First World War* (Stroud: The History Press, 2013), 20.
49. Hallett, *Veiled Warriors*, 68.
50. Powell, *Women in the War Zone*, 129.
51. *A War Nurse's Diary: Sketches from a Belgian Field Hospital* (New York: The Macmillan Company, 1918), 14–15.
52. Hallett, *Veiled Warriors*, 244.
53. "Memorandum: To all Medical Officers, 32d Division, April 11, 1918" (n.p.). Clendening History of Medicine Museum, Kansas City, Missouri.
54. Michael B. Tyquin, *Gallipoli: The Medical War: The Australian Army Medical Services in the Dardanelles Campaign of 1915* (Kensington: New South Wales University Press, 1993), 53.
55. *The War on Hospital Ships: From the Narratives of Eye-Witnesses* (London: T. Fisher Unwin, Ltd., 1917), 5.
56. *The War on Hospital Ships*, 6.

57. John M. Barry, "Journal of the Plague Year," *Smithsonian*, 48(7) November 2017, 36.
58. John M. Barry, *The Great Influenza: The Story of the Deadliest Pandemic in History* (New York: Penguin Books, 2005), 173.
59. Barry, *The Great Influenza*, 372.
60. "Epidemic of Influenza." Letter by H. A. May quoted in Sarah Sand Stevenson's *Lamp for a Soldier: The Caring Story of a Nurse in World War I* (North Dakota: North Dakota State Nurses' Association, 1976), 35.
61. Barry, *The Great Influenza*, 370–71.
62. Francisco, *Wartime Diary*.
63. Francisco, *Wartime Diary*.
64. Francisco, *Wartime Diary*.
65. Francisco, *Wartime Diary*.
66. "Rise of Infant Mortality as a Consequence of War," *American Medical Association Journal* 65 (3 July 1915), 41.
67. Nancy K. Bristow, *American Pandemic: The Lost Worlds of the 1918 Influenza Epidemic* (New York: Oxford University Press, 2012), 137–38.
68. W. Douglas Fisher and Joann H. Buckley, *African American Doctors of World War I: The Lives of 104 Volunteers* (Jefferson: McFarland & Company, Inc., Publishers, 2016), 13.
69. Whitehead, *Doctors in War*, 260.
70. F.N.L. Poynter, ed., *Medicine and Surgery in the Great War, 1914–1918* (London: The Wellcome Institute of the History of Medicine at War, 1968), 6–7.
71. Richard V.M. Ginn, *The History of the U.S. Army Medical Service Corps* (Washington, DC: Office of the Surgeon General and Center of Military History United States Army, 1997), 37.
72. *Base Hospital No. 52 War Diary* (United States: n.p., 1919), 1.
73. "April 15, 1918 Memorandum."

4 Soldiers and the Medical Front

1. Ruth Cowen, ed., *A Nurse at the Front: The First World War Diaries of Sister Edith Appleton* (London: Simon & Schuster, 2012), 228.
2. Will R. Bird, *And We Go On: A Memoir of the Great War* (Montreal: McGill-Queen's University Press, 2014), 22–23.
3. Harvey Cushing, *From a Surgeon's Journal, 1915–1918* (Boston: Little, Brown, and Company, 1936), 29.
4. Robert V. Dolbey, *A Regimental Surgeon in War and Peace* (London: John Murray, 1917), 94.

5. H. H. Storm, *A Soldier's Diary of World War One, France 1917–1919* (Rockport, ME: Lynwood Editions, 2006), 43.
6. Rachel Duffett, "Ingestion and Digestion on the Western Front," in *Modern Conflict and the Senses*, eds. Nicholas J. Saunders and Paul Cornish (New York: Routledge, 2017), 171–72.
7. Duffett, "Ingestion and Digestion," 172.
8. Harry Stinton, *Harry's War: Experiences in the "Suicide Club" in World War One* (London: Brassey's, 2002), 76.
9. Stinton, *Harry's War*, 210.
10. J. Garry Clifford, ed., *The World War I Memoirs of Robert P. Patterson: A Captain in the Great War* (Knoxville: The University of Tennessee Press, 2012), 60.
11. John Lewis Barkley, *Scarlet Fields: The Combat Memoir of a World War I Medal of Honor Hero* (Lawrence: The University Press of Kansas, 2012), 194–95.
12. "April 15, 1918, Memorandum" (n.p.).
13. Cushing, *From a Surgeon's Journal*, 44–45.
14. Cushing, *From a Surgeon's Journal*, 52.
15. *History of Base Hospital No. 59* (United States: n.p., 1919), 8.
16. Thomas Scotland and Steven Heys, eds., *War Surgery, 1914–18* (West Midlands: Helion & Company, 2013), 223.
17. Storm, *A Soldier's Diary*, 89.
18. Barkley, *Scarlet Fields*, 101.
19. Dolbey, *A Regimental Surgeon*, 95.
20. Arthur Anderson Martin, *A Surgeon in Khaki* (London: Edward Arnold, 1915), 62–64.
21. Clifford, *The World War I Memoirs of Robert P. Patterson*, 39.
22. H. Winnett Orr, *An Orthopedic Surgeon's Story of the Great War* (Lincoln: n.p., 1921), 26.
23. Tim Cook, *No Place to Run: The Canadian Corps and Gas Warfare in the First World War* (Vancouver: UBC Press, 2013), 163–64.
24. Clifford, 45.
25. Barkley, *Scarlet Fields*, 153–54.
26. Barkley, *Scarlet Fields*, 181.
27. Barkley, *Scarlet Fields*, 218.
28. Fiona Reid, *Medicine in First World War Europe: Soldiers, Medics, Pacifists* (London: Bloomsbury, 2017), 116–17.
29. Dolbey, *A Regimental Surgeon*, 34–35.
30. Barkley, *Scarlet Fields*, 84–90.
31. Clifford, *The World War I Memoirs of Robert P. Patterson*, 21.
32. Barkley, *Scarlet Fields*, 106–7, 118.
33. Barkley, *Scarlet Fields*, 206–7.

34. Allan Kent Powell, ed., *Nels Anderson's World War I Diary* (Salt Lake City: The University of Utah Press, 2013), 111.
35. Powell, *Nels Anderson's World War I Diary*, 127.
36. Powell, *Nels Anderson's World War I Diary*, 134.
37. Clifford, *The World War I Memoirs of Robert P. Patterson*, 78.
38. Barkley, *Scarlet Fields*, 224.
39. Dolbey, *A Regimental Surgeon*, 75.
40. Powell, *Nels Anderson's World War I Diary*, 129.
41. Powell, *Nels Anderson's World War I Diary*, 131.
42. Powell, *Nels Anderson's World War I Diary*, 95.
43. Powell, *Nels Anderson's World War I Diary*, 115.
44. Barkley, *Scarlet Fields*, 75.
45. Barkley, *Scarlet Fields*, 120.
46. *Base Hospital No. 59* (US: n.p., 1919), 6.
47. *Base Hospital No. 59* (US: n.p., 1919), 6.
48. Clifford, *The World War I Memoirs of Robert P. Patterson*, 17.
49. Marjorie Barron Norris, ed., *Medicine and Duty: The World War I Memoir of Captain Harold W. McGill, Medical Officer, 31st Battalion C.E.F.* (Calgary: University of Calgary Press, 2007), 49.
50. Clifford, 27.
51. Barkley, *Scarlet Fields*, 30.
52. *Smash the Line!* (New York: War Department, Commission on Training Camp Activities, 1917), 1.
53. Storm, *A Soldier's Diary*, 51–52.
54. *Base Hospital No. 59*, 4.
55. Leo Mayer, *The Military Orthopedic Reconstruction Hospital* (Chicago: The American Medical Association, 1917), 7.
56. Mayer, *The Military Orthopedic Reconstruction Hospital*, 4.
57. Robert B. Osgood, *A Survey of the Orthopaedic Services in the U.S. Army Hospitals, General, Base, and Debarkation* (Boston: Jamaica Printing, 1919), 6.
58. *Home Service and the Disabled Soldier or Sailor* (Washington, DC: The American Red Cross, 1918), 12.
59. Jeffrey S. Reznick, *Healing the Nation: Soldiers and the Culture of Caregiving in Britain during the Great War* (Manchester: Manchester University Press, 2004), 66.
60. Barkley, *Scarlet Fields*, 95–96.
61. Powell, *Nels Anderson's World War I Diary*, 108.
62. Clifford, *The World War I Memoirs of Robert P. Patterson*, 76.
63. Scotland and Heyes, *War Surgery, 1914–18*, 206.
64. Clifford, *The World War I Memoirs of Robert P. Patterson*, 52.

5 Effects of the Medical Front in the Great War

1. James H. Nicoll, "The Femoral Artery in War Surgery," *British Medical Journal* 2 (November 23, 1918), 570.
2. Dafydd S. Edwards, Emily R. Mayhew, and Andres S. C. Rice, "'Doomed to Go in Company with Miserable Pain': Surgical Recognition and Treatment of Amputation-Related Pain on the Western Front during World War 1," *The Lancet* 384 (November 8, 2014), 1716.
3. Ian R. Whitehead, *Doctors in War* (Barnsley: Pen & Sword Military, 1998), 258.
4. William H. Schneider, "Blood Transfusion in Peace and War, 1900–1918," *The Society for the Social History of Medicine* 10 (April 1997), 126.
5. Harvey Cushing, *From a Surgeon's Journal, 1915–1918* (Boston: Little, Brown, and Company, 1936), 495.
6. Michael Bliss, *Harvey Cushing: A Life in Surgery* (New York: Oxford University Press, 2005), 282–83.
7. Bliss, *Harvey Cushing*, 292.
8. Barbara Goldsmith, *Obsessive Genius: The Inner World of Marie Curie* (New York: W.W. Norton, 2005), 186–89.
9. *United States Army X-Ray Manual* (New York: Paul B. Hoeber, 1919), 476.
10. Julian W. Mack, "A Chance—With a Running Start: Government Compensation Provides Means for the Handicapped Fighter," *Carry On*, 1, no. 2 (August 1918), 11.
11. "An Embryo Gardener," *Carry On*, 1, no.1 (June 1918), 16.
12. Beth Linker, *War's Waste: Rehabilitation in World War I America* (Chicago: The University of Chicago Press, 2011), 7.
13. Linker, *War's Waste*, 59.
14. Linker, *War's Waste*, 59.
15. Jeffrey S. Reznick, "Prostheses and Propaganda," in *Matters of Conflict: Material Culture, Memory, and the First World War*, ed. Nicholas J. Saunders (London: Routledge, 2004), 54–58.
16. J. Garry Clifford, ed., *The World War I Memoirs of Robert P. Patterson: A Captain in the Great War* (Knoxville: The University of Tennessee Press, 2012), 100.
17. Charles Lynch, Frank W. Weed, and Loy McAfee, eds., *The Medical Department of the United States Army in the World War, Surgery*, vol. 11, pt. 1 (Washington, DC: Government Printing Office, 1927), 552.
18. Linker, *War's Waste*, 80.
19. Merritte W. Ireland, *The Achievement of the Army Medical Department in the World War: In Light of General Medical Progress* (Chicago: American Medical Association, 1921), 14.

20. Douglas C. McMurtrie, *The Duty of the Medical Profession in the Reconstruction of the War Cripple* (New York: The American Red Cross, 1918), 1.
21. Reznick, "Prostheses and Propaganda," 59.
22. Fiona Reid, *Broken Men: Shell Shock, Treatment and Recovery in Britain, 1914–1930* (London: Bloomsbury, 2014), 100.
23. Richard A. Gabriel and Karen S. Metz, *A History of Military Medicine: Volume II, From the Renaissance through Modern Times* (New York: Greenwood Press, 1992), 241.
24. "French Orthopedic Society," *American Medical Association Journal* 71 (November 30, 1918), 1842.
25. Linker, *War's Waste*, 35.
26. "History," The American Orthopaedic Association, n.d., http://www.aoassn.org/aoaimis/AOANEW/About/History/AOANEW/About_AOA/History_Founding.aspx?hkey=026d8c93-474b-4fa2-acaf-6b9e154fc8c1.
27. Roger Cooter, *Surgery and Society in Peace and War: Orthopaedics and the Organization of Modern Medicine, 1880–1948* (London: Macmillan Press, 1993), 136.
28. Robert B. Osgood, *A Survey of the Orthopaedic Services in the U.S. Army Hospitals, General, Base, and Debarkation* (Boston: Jamaica Printing, 1919), 5.
29. Osgood, *A Survey of the Orthopaedic Services*, 14.
30. Fred H. Albee, *The Function of the Military Orthopedic Hospital* (New York: A.R. Elliott Publishing, 1917), 5–7.
31. J. Trueta, *Treatment of War Wounds and Fractures with Special Reference to the Closed Method as Used in the War in Spain* (London: Hamish Hamilton Medical Books, 1939), 10.
32. Murray C. Meikle, *Reconstructing Faces: The Art and Wartime Surgery of Gillies, Pickerill, McIndoe & Mowlem* (Otago: Otago University Press, 2013), 37.
33. Anthony F. Wallace, *The Progress of Plastic Surgery: An Introductory History* (Oxford: William A. Meeuws, 1982), 39.
34. Wallace, *The Progress of Plastic Surgery*, 40.
35. Wallace, *The Progress of Plastic Surgery*, 40.
36. Wallace, *The Progress of Plastic Surgery*, 150–51.
37. Fred. H. Albee, *Restoration of Loss of Bone: Including an Analysis of the First Hundred Cases of Fracture Treated by Bone Graft at U.S. Army General Hospital No. 3, Colonia, N.J.* (Chicago: The American Medical Association, 1920), 4–6.
38. Wallace, *The Progress of Plastic Surgery*, 170.
39. Meikle, *Reconstructing Faces*, 116.
40. Wallace, *The Progress of Plastic Surgery*, 171.
41. Meikle, *Reconstructing Faces*, 116–17.
42. Meikle, *Reconstructing Faces*, 113.
43. Roy Porter, *The Greatest Benefit to Mankind: A Medical History of Humanity* (New York: W.W. Norton, 1997), 637.
44. Meikle, *Reconstructing Faces*, 120.

45. Meikle, *Reconstructing Faces*, 120.
46. Wallace, *The Progress of Plastic Surgery*, 71, 139.
47. J.M.T. Finney quoted in Merritte W. Ireland, *The Achievement of the Army Medical Department in the World War: In Light of General Medical Progress* (Chicago: American Medical Association, 1921), 5.
48. Schneider, 125.
49. John Pearn, "Civilian Legacies of Army Health," in *Health and History* 6, no. 2 (2004), 13.
50. J.M. Weddell, "The Treatment of Wounds in War," in *British Medical Journal* 1 (April 15, 1939), 787.
51. Weddell, "The Treatment of Wounds," 785–87.
52. Thomas Scotland and Steven Heys, eds., *War Surgery 1914–18* (Solihull: Helion & Company, 2013), 204.
53. Scotland and Heys, *War Surgery*, 238.
54. Scotland and Heys, *War Surgery*, 245.
55. Scotland and Heys, *War Surgery*, 245.
56. Nancy K. Bristow, *American Pandemic: The Lost Worlds of the 1918 Influenza Epidemic* (New York: Oxford University Press, 2012), 89.
57. Bristow, *American Pandemic*, 90.
58. Alexandra Minna Stern, Martin S. Centron, and Howard Markel, "The 1918–1919 Influenza Pandemic in the United States: Lessons Learned and Challenges Exposed," *Public Health Reports* 125, Supplement 3 (2010), 6.
59. G. Dennis Shanks, "How World War 1 Changed Global Attitudes to War and Infectious Diseases," *Lancet* 384 (2014), 1701.
60. Shanks, "How World War 1 Changed Global Attitudes," 1700.
61. Peter Cornelis Wever and Leo van Bergen, "Prevention of Tetanus during the First World War," *Journal of Medical Humanities* 38 (2012), 80.
62. Wever and van Bergen, "Prevention of Tetanus," 81.
63. Wever and van Bergen, "Prevention of Tetanus," 82.
64. Porter, *The Greatest Benefit*, 642.
65. Porter, *The Greatest Benefit*, 642.
66. Porter, *The Greatest Benefit*, 642.
67. Richard V.N. Ginn, *The History of the U.S. Army Medical Service Corps* (Washington, DC: Office of the Surgeon General and Center of Military History United States Army, 1997), 81.
68. Porter, *The Greatest Benefit*, 399.
69. Ireland, *The Achievement of the Army Medical*, 14.
70. Shanks, "How World War 1 Changed Global Attitudes," 1700.
71. Ireland, *The Achievement of the Army Medical*, 14.
72. John C. Burnham, *Health Care in America: A History* (Baltimore: Johns Hopkins University Press, 2015), 244.
73. Porter, *The Greatest Benefit*, 642.

74. Charles H. Mayo, *Educational Possibilities of the National Medical Museum in the Standardization of Medical Training* (Chicago: The American Medical Association, 1919), 7.
75. Mayo, *Educational Possibilities*, 4.
76. Mayo, *Educational Possibilities*, 7.
77. Ireland, *The Achievement of the Army Medical*, 18.
78. Burnham, *Health Care in America*, 249–50.
79. Edwards, Mayhew, and Rice, "Doomed to Go," 1715.
80. Edwards, Mayhew, and Rice, "Doomed to Go," 1717.
81. Meickle, 64.
82. William H. Schneider, "Blood Transfusion in Peace and War, 1900–1918," *The Society for the Social History of Medicine* 10 (April 1997), 126.
83. Roger Cooter and Steve Sturdy, "Of War, Medicine and Modernity: Introduction," in *War, Medicine and Modernity*, eds. Roger Cooter, Mark Harrison, and Steve Sturdy (Thrupp: Sutton Publishing, 1998), 15.
84. Ireland, *The Achievement of the Army Medical*, 7.
85. "The Medical Profession After the War," *American Medical Association Journal* 71 (November 16, 1918), 1663.
86. Burnham, *Health Care in America*, 209.
87. Cooter and Sturdy, "Of War, Medicine and Modernity," 12.
88. Ireland, *The Achievement of the Army Medical*, 16.
89. Weddell, "The Treatment of Wounds in War," 785.
90. Ireland, *The Achievement of the Army Medical*, 17.
91. Mayo, *Educational Possibilities*, 10.
92. Mayo, *Educational Possibilities*, 4.
93. Mayo, *Educational Possibilities*, 5.
94. H. Winnett Orr, *An Orthopedic Surgeon's Story of the Great War* (Lincoln: n.p., 1921), Introduction, 13.

Conclusion

1. Charles H. Mayo, *Educational Possibilities of the National Medical Museum in the Standardization of Medical Training* (Chicago: The American Medical Association, 1919), 4.
2. Gordon Corrigan, *Mud, Blood and Poppycock: Britain and the First World War* (London: Cassell, 2003), 401.
3. Dan Todman, *The Great War: Myth and Memory* (London: Hambledon and London, 2005), 177–78.
4. Roy Porter, *The Greatest Benefit to Mankind: A Medical History of Humanity* (New York: W.W. Norton & Company, 1997), 642.

Acknowledgments

The Great War has been everywhere in the past few years. The centenary of the war began in 2014. There have been new museum exhibits and popular media news stories; dramatizations have appeared, and, I hope, there has been in the public's mind a rekindling of interest in this conflict. In the United States, World War I has often been shunted aside. The United States did not participate in the war for very long, and it did not suffer in the same way as the countrymen of Europe did. The history texts skim over it to arrive at the more popular topic of the Second World War. Each time this happens, a great disservice is done. I firmly believe that one cannot understand the rest of the twentieth century without understanding the First World War. My hope is that even after the fervor of the centenary passes, people will still see the merit in studying and grasping the impact of the Great War.

The war is often painted in broad strokes and catchphrases meant to sum up the conflict—trench warfare, U-boats, poison gas, "the war to end all wars." What gets lost in these descriptions is the war at a personal level. One place to find that humanity and what the impact was on the individual is with the study of the medical side of the conflict. A person can become acquainted not only with the medical innovations but with the people who delivered them and the people who benefited from them.

Acknowledgments

The medical aspects of this conflict have occupied much of my life for the past several years as I worked on this book. I would like to extend my gratitude to a number of individuals and institutions who aided me in the writing of this work: first, Professor Arnold Krammer, in whose class in graduate school I first became truly exposed to World War I and who supported my research into the medical exploits of it in a seminar paper. I had never read anything in depth on the war, and I immediately became both horrified and mesmerized by it. I dove into reading everything about it I could get my hands on, and my fascination with it never left.

As this project began to solidify, I made several trips, doing research for this book, and I cannot thank everyone who personally assisted me, but there are some particular institutions of which I need to make mention. The National World War I Museum in Kansas City, Missouri, had a number of items in its archives that I made use of on a research trip. Jonathan Casey, his staff, and volunteers were exceedingly helpful during my days of research there. The staff of the Clendening History of Medicine Museum at the University of Kansas Medical Center was also extremely helpful and made copies of anything I asked for and provided a very quiet room for research. Dr. Frederick Holmes assisted in getting me acclimated at the Clendening. The Wellcome Unit for the History of Medicine at Oxford University had the most wonderful focused library to utilize, and I appreciate Mark Harrison taking the time to meet with me when I was there.

The US National Library of Medicine has a robust collection of World War I materials. Many of these have been digitized and are readily available to the public through its website. These resources were invaluable as I am not located near any of these previously mentioned archives. The Interlibrary Loan office at the University of Texas at Arlington and the librarians at Tarrant County College always diligently obtained the materials I would request. Special thanks especially to the staff of the Judith C. Carrier Library at TCC's Southeast Campus, who found ways to make sure I could check out any number of items even if they exceeded the system's limits. I would also like to thank my editor, Rachel Bridgewater, and the staff at Palgrave that I worked with on this project. My friends and colleagues have always warmly encouraged this project, and I have appreciated their support and kind words.

My family has also been unflaggingly supportive of this book project. My husband, Hoyt, encouraged me to dive back into the research for this book and to stay focused on it through the months and years it took to come to fruition despite my many attempts to distract myself with, well, anything else. My parents too always cheered me on through the years of school, school, and more school. My mother was always interested to hear about how the work was progressing. I learned to love history at a young age from my father, who taught history at the local college in my hometown. I blame him for the bevy of books that have inundated my life over the years. My cat will be quite pleased that this book is finished, as she did not appreciate being ignored in favor of work. I also extend a special and heartfelt thanks for all the people in my life who have put up with my World War I stories through the years with such good humor.

Bibliography

Primary Sources

Albee, Fred. H. *The Function of the Military Orthopedic Hospital*. New York: A.R. Elliott, 1917.

——. *Restoration of Loss of Bone: Including an Analysis of the First Hundred Cases of Fracture Treated by Bone Graft at U.S. Army General Hospital No. 3, Colonia, N.J.* Chicago: The American Medical Association, 1920.

Barkley, John Lewis. *Scarlet Fields: The Combat Memoir of a World War I Medal of Honor Hero*. Lawrence: The University Press of Kansas, 2012.

Base Hospital No. 52 War Diary. United States: n.p., 1919. Available through the US National Library of Medicine, Bethesda, MD.

Bird, Will R. *And We Go On: A Memoir of the Great War*. Montreal: McGill-Queen's University Press, 2014.

Blake, Joseph A. *Gun-Shot Fractures of the Extremities*. Paris: Masson et Cie, Éditeurs, 1918.

"Blood Transfusion in War Surgery." *American Medical Association Journal* 70, no. 5 (January 1918): 59–60.

Cheatle, G. Lenthal. "Antiseptics in War." *British Medical Journal* 2 (December 12, 1914), 1006.

Clifford, J. Garry, ed. *The World War I Memoirs of Robert P. Patterson: A Captain in the Great War*. Knoxville: University of Tennessee Press, 2012.

Cowen, Ruth Cowen, ed. *A Nurse at the Front: The Great War Diaries of Sister Edith Appleton*. London: Simon & Schuster, 2012.

Cushing, Harvey. *From a Surgeon's Journal, 1915–1918.* Boston: Little, Brown, 1936.
Dolbey, Robert V. *A Regimental Surgeon in War and Peace.* London: John Murray, 1917.
Dorsett, Rae S. *History of Base Hospital No. 85, United States Army.* United States: n.p., 1919. Available through the US National Library of Medicine, Bethesda, MD.
Don, A. "Dressings Used in a Casualty Clearing Station." *British Medical Journal* 1 (May 6, 1916), 648–49.
Drennen, W. Earle. "Experiences in Military Surgery." *American Medical Association Journal* 65 (July 24, 1915): 296–300.
"Early Suture of War Wounds." *American Medical Association Journal* 70 (January 5, 1918): 61.
"An Embryo Gardener." *Carry On* 1, no.1 (June 1918): 16.
Fauntleroy, A.M. *Report on the Medico-Military Aspects of the European War.* Washington, DC: Government Printing Office, 1916.
Fisher, A. R. "Chloramine in the Treatment of Wounds of the Mouth and Jaws." *British Medical Journal* 1 (January 15, 1916), 87–88.
Francisco, Clarence Benjamin. *Wartime Diary of Clarence Benjamin Francisco, M.D.* n.p. Available through the Clendening History of Medicine Museum, Kansas City, Missouri.
"French Orthopedic Society." *American Medical Association Journal* 71 (November 30, 1918): 1842.
"German and British Bullets." *British Medical Journal* 1 (December 12, 1914): 1041.
"German, French, and British Bullets." *British Medical Journal* 1 (December 5, 1914): 99–91.
Goodwin, T. H. *Notes for Army Medical Officers.* Philadelphia: Lea & Febiger, 1917.
Groves, Hey. "An Address on Some of the Principles and Problems Related to the Treatment of Gunshot Fractures." *British Medical Journal* 2 (July 15, 1916): 65–70.
Hanson, William L. *World War I: I Was There.* Gerald: The Patrice Press, 1982.
History of Base Hospital No. 59. United States: n.p. 1919. Available through the US National Library of Medicine, Bethesda, MD.
Home Service and the Disabled Soldier or Sailor. Washington, DC: The American Red Cross, 1918.
Horton, Charles H. *Stretcher Bearer! Fighting for Life in the Trenches*, edited by Dale Le Vack. Oxford: Lion Books, 2013.

Ireland, Merritte W. *The Achievement of the Army Medical Department in the World War: In Light of General Medical Progress.* Chicago: American Medical Association, 1921.

Jones, Robert. "An Address on the Orthopaedic Outlook in Military Surgery." *British Medical Journal* 1 (January 12, 1918): 41–45.

Kazanjian, V. H. "Treatment of Maxillary Fractures." *British Medical Journal* 1 (February 19, 1916): 266–67.

LaMotte, Ellen N. *The Backwash of War: The Human Wreckage of the Battlefield as Witnessed by an American Hospital Nurse.* New York: G.P. Putnam's Sons, 1916.

Lynch, Charles, Frank W. Weed, and Loy McAfee, eds. *The Medical Department of the United States Army in the World War*, Vols. 1–15. Washington: US Army Surgeon General's Office, 1923–29.

Mack, Julian W. "A Chance—With a Running Start: Government Compensation Provides Means for the Handicapped Fighter." *Carry On* 1, no. 2 (August 1918): 11–13.

Martin, Arthur Anderson. *A Surgeon in Khaki.* London: Edward Arnold, 1915.

Mayer, Leo. *The Military Orthopedic Reconstruction Hospital.* Chicago: American Medical Association, 1917.

Mayo, Charles H. *Educational Possibilities of the National Medical Museum in the Standardization of Medical Training.* Chicago: American Medical Association, 1919.

McMurtrie, Douglas C. McMurtrie. *The Duty of the Medical Profession in the Reconstruction of the War Cripple.* New York: American Red Cross, 1918.

"The Medical Profession After the War." *American Medical Association Journal* 71 (November 16, 1918): 1663–64.

Medical Training Camp, Fort Oglethorpe, Georgia, August 1917. n.p., 1917.

"Memorandum: To all Medical Officers, 32d Division, April 11, 1918." n.p. Clendening History of Medicine Museum, Kansas City, Missouri.

"April 15, 1918, Memorandum." n.p. Clendening History of Medicine Museum, Kansas City, Missouri.

Nicoll, James H. Nicoll. "The Femoral Artery in War Surgery." *British Medical Journal* 2 (November 23, 1918): 569–70.

"The Femoral Artery in War Surgery." *British Medical Journal* 2 (November 23, 1918): 569–70.

Norris, M. Barron, ed. *Medicine and Duty: The World War I Memoir of Captain Harold W. McGill, Medical Officer, 31st Battalion C.E.F.* Calgary: University of Calgary Press, 2007.

Orr, H. Winnett. *An Orthopedic Surgeon's Story of the Great War*. Lincoln: n.p., 1921.
Pottle, Frederick A. *Stretchers: The Story of a Hospital Unit on the Western Front*. New Haven: Yale University Press, 1929.
Powell, Allan Kent, ed. *Nels Anderson's World War I Diary*. Salt Lake City: University of Utah Press, 2013.
"Primary Suture of Wounds." *American Medical Association Journal* 65 (July 17, 1915): 228.
"Radiography in the Diagnosis of Bullet Wounds." *British Medical Journal* 1 (December 12, 1914): 1047.
"Results Obtained at an Advanced Surgical Unit." *American Medical Association Journal* 70 (January 12, 1918): 111.
Richet, Charles. *War Nursing: What Every Woman Should Know, Red Cross Lectures*. transl. Helen de Vere Beauclerk. New York: Robert M. McBride, 1918.
"Rise in Infant Mortality as a Consequence of War." *American Medical Association Journal* 65 (July 3, 1915): 41.
Schneider, William H. "Blood Transfusion in Peace and War, 1900–1918." *Social History of Medicine* 10 (April 1997): 105–26.
Smash the Line! New York: War Department, Commission on Training Camp Activities, 1917.
Starr, M. A. "Sculptor's Art in Surgery; A Branch of War Medicine That We Must Consider in Our Turn." *New York Times*, July 3, 1918, 12.
Stevenson, Sarah Sand. *Lamp for a Soldier: The Caring Story of a Nurse in World War I*. Bismarck: North Dakota State Nurses' Association, 1976.
Stinton, Harry. *Harry's War: Experiences in the "Suicide Club" in World War One*. London: Brassey's, 2002.
Storm, H. H. *A Soldier's Diary of World War One, France 1917–1919*. Rockport, ME: Lynwood Editions, 2006.
"Treatment of Projectile Wounds by Excision of Damaged Tissues." *American Medical Association Journal* 65 (31 July 1915): 463.
"Treatment of War Wounds." *American Medical Association Journal* 70 (19 January 1918): 195.
United States Army X-Ray Manual. New York: Paul B. Hoeber, 1919.
The War on Hospital Ships: From the Narratives of Eye-witnesses. London: T. Fisher Unwin, 1917.
"War Orthopaedics." *British Medical Journal* 2 (26 October 1915): 575–76.
A War Nurse's Diary: Sketches from a Belgian Field Hospital. New York: Macmillan Company, 1918.
"Wounds of Arteries in Legs." *American Medical Association Journal* 70 (12 January 1918): 133.

Secondary Sources

Barry, John M. *The Great Influenza: The Story of the Deadliest Pandemic in History*. New York: Penguin Books, 2005.
Bliss, Michael. *Harvey Cushing: A Life in Surgery*. New York: Oxford University Press, 2005.
Bristow, Nancy K. *American Pandemic: The Lost Worlds of the 1918 Influenza Epidemic*. New York: Oxford University Press, 2012.
Burnham, John C. *Health Care in America: A History*. Baltimore: Johns Hopkins University Press, 2015.
Clodfelter, Michael. *Warfare and Armed Conflicts: A Statistical Reference to Casualty and Other Figures, 1618–1991*. Jefferson: McFarland, 1992.
Cook, Tim. *No Place to Run: The Canadian Corps and Gas Warfare in the First World War*. Vancouver: UBC Press, 2013.
Cooter, Roger. *Surgery and Society in Peace and War: Orthopaedics and the Organization of Modern Medicine, 1880–1948*. London: Macmillan Press, 1993.
Cooter, Roger, Mark Harrison, and Steve Sturdy, eds. *War, Medicine and Modernity*. Thrupp: Sutton Publishing, 1998.
Corrigan, Gordon. *Mud, Blood and Poppycock: Britain and the First World War*. London: Cassell, 2003.
Duffet, Rachel. "Ingestion and Digestion on the Western Front." In *Modern Conflict and the Senses*, eds. Nicholas J. Saunders and Paul Cornish. New York: Routledge, 2017.
Emery, Theo. *Hellfire Boys: The Birth of the U.S. Chemical Warfare Service and the Race for the World's Deadliest Weapons*. New York: Little, Brown, 2017.
Emling, Shelley. *Marie Curie and Her Daughters: The Private Lives of Science's First Family*. New York: Palgrave Macmillan, 2012.
Fisher, W. Douglas, and Joann H. Buckley. *African American Doctors of World War I: The Lives of 104 Volunteers*. Jefferson: McFarland & Company, Inc., Publishers, 2016.
Gabriel, Richard A., and Karen S. Metz. *A History of Military Medicine: Volume II, From the Renaissance through Modern Times*. New York: Greenwood Press, 1992.
Ginn, Richard V.M. *The History of the U.S. Army Medical Service Corps*. Washington, DC: Office of the Surgeon General and Center of Military History United States Army, 1997.
Goldsmith, Barbara. *Obsessive Genius: The Inner World of Marie Curie*. New York: W.W. Norton & Company, 2005.
Hallett, Christine. *Containing Trauma: Nursing Work in the First World War*. Manchester: Manchester University Press, 2009.

Hallett, Christine. *Veiled Warriors: Allied Nurses of the First World War*. New York: Oxford University Press, 2014.
Harrison, Mark. *The Medical War: British Military Medicine in the First World War*. Oxford: Oxford University Press, 2010.
Mark Harrison, "The Medicalization of War—The Militarization of Medicine." *Journal of the Social History of Medicine* 9 (August 1996), 267–68.
Linker, Beth. *War's Waste: Rehabilitation in World War I America*. Chicago: The University of Chicago Press, 2011.
Major, Ralph H. *A History of Medicine*, Vol. 2. Springfield: Charles C. Thomas Publishers, 1954.
Mayhew, Emily. *Wounded: A New History of the Western Front in World War I*. New York: Oxford University Press, 2013.
McMeekin, Sean. *July 1914: Countdown to War*. New York: Basic Books, 2013.
Meikle, Murray C. *Reconstructing Faces: The Art and Wartime Surgery of Gillies, Pickerill, McIndoe & Mowlem*. Otago: Otago University Press, 2013.
Naythons, Matthew. *Faces of Mercy: A Photographic History of Medicine at War*. New York: Random House, 1993.
Osgood, Robert B. *A Survey of the Orthopaedic Services in the U.S. Army Hospitals, General, Base, and Debarkation*. Boston: Jamaica Printing, 1919.
Porter, Roy. *The Greatest Benefit to Mankind: A Medical History of Humanity*. New York: W.W. Norton, 1997.
Powell, Anne. *Women in the War Zone: Hospital Service in the First World War*. Stroud: The History Press, 2013.
Poynter, F.N.L., ed. *Medicine and Surgery in the Great War, 1914–1918*. London: The Wellcome Institute of the History of Medicine at War, 1968.
Reid, Fiona. *Broken Men: Shell Shock, Treatment and Recovery in Britain, 1914–1930*. London: Bloomsbury, 2014.
Reid, Fiona. *Medicine in First World War Europe: Soldiers, Medics, Pacifists*. London: Bloomsbury, 2017.
Reznick, Jeffrey S. "Prostheses and Propaganda" in *Matters of Conflict: Material Culture, Memory, and the First World War*, ed. Nicholas J. Saunders. London: Routledge, 2004.
Scotland, Thomas and Steven Heys, eds. *War Surgery, 1914–1918*. West Midlands: Helion & Company, Ltd., 2013.
Stern, Alexandra Minna, Martin S. Centron, and Howard Markel, "The 1918–1919 Influenza Pandemic in the United States: Lessons Learned and Challenges Exposed." *Public Health Reports* 125, Supplement 3 (2010), 6–8.
Strachan, Hew. *The First World War*. New York: Viking, 2003.

Todman, Dan. *The Great War: Myth and Memory*. London: Hambledon and London, 2005.
Trueta, J. *Treatment of War Wounds and Fractures with Special Reference to the Closed Method as Used in the War in Spain*. London: Hamish Hamilton Medical Books, 1939.
Tyquin, Michael B. *Gallipoli: The Medical War: The Australian Army Medical Services in the Dardanelles Campaign of 1915*. Kensington: New South Wales University Press, 1993.
Van Bergen, Leo. *Before My Helpless Sight: Suffering, Dying and Military Medicine on the Western Front, 1914–1918*. Surrey: Ashgate, 2009.
Van Tiggelen, René. *Radiology in a Trench Coat: Military Radiology on the Western Front during the Great War*. Brussels: n.p., 2013.
Wallace, Anthony F. *The Progress of Plastic Surgery: An Introductory History*. Oxford: William A. Meeuws, 1982.
Wever, Peter Cornelis, and Leo van Bergen, "Prevention of Tetanus during the First World War." *Journal of Medical Humanities* 38 (2012), 78–82.
Whitehead, Ian R. *Doctors in War*. Barnsley: Pen & Sword Military, 1998.

Journals

American Medical Association Journal
Beam Line
British Journal for the History of Science
British Medical Journal
Carry On
Health and History
Journal of Medical Humanities
Journal of the Social History of Medicine
Lancet
New York Times
Public Health Reports
Smithsonian
Transfusion Medicine

Index

A

Aftercare, 102–103
  frustration with, 114
Airplanes, 16
Ambulance, 87–88
  corps, 27
  mobile unit, 89
American Orthopedic Association, 65
Amputation, 12, 33–36, 49, 91, 102, 114, 117, 132
Anderson, Nels, 96, 98
Appleton, Edith, 71
Armistice, 10
Austro-Hungary, 3–5, 10

B

Bacteriology, 34
Barkley, John Lewis, 93–97, 100, 105, 133
Base hospitals, 26, 28, 98–99
Base Hospital No.52, 81

Belgium, invasion of, 6
Bird, Will R., 21, 85
Blood transfusion, 38–40, 106, 123
Blood typing, 39, 123
Bone grafts, 119–120
Bullets, 15, 29–30
Bureaucracy, 26, 50, 56, 68, 80–81, 86, 126–128

C

Carrel, Alexis, 35
Carrel-Dakin method, 31–33, 122
Carry On, 112–113
Casualty-clearing stations, 26, 28, 40, 66, 70, 94, 99
Civilian medicine, benefits to, xiv, 25, 12, 123–125, 135
Civilians
  cases, 72
  treatment of in warzone, 58
  treatment of home-front, 78
Crile, George, 13

171

Concussions, 125
Corrigan, Gordon, 141
Crowding, 99–100
Curie, Marie, 44, 112
Cushing, Harvey, 110–111, 125

Debridement, 32–33, 122, 132, 134
Demobilization, 104, 109–110, 115
Dental surgeons, 49, 120
Doctors, journals, 58–59
Dolbey, Robert V., 57

Evacuation, 56, 67, 85, 88
Evacuation hospitals (U.S.), 27–28, 66–67, 70
Ex-Services Welfare Society (ESWS), 115

Facial masks, 49, 133
Field Ambulance, 26–27, 34, 57, 90–91
Francisco, Benjamin, 55–56, 77–78
Franz Ferdinand, xiii, 3–4

Gas gangrene, 29, 32–33, 45, 78, 84, 131–132
Gas masks, 18, 70, 92–93
Germ theory of disease, 13, 30, 127
German prisoners, 65, 94–95

Gillies, Harold Delf, 120–121
Government, expectations of, 138, 140–141

Hallett, Christine, xvii–xviii, 72
Hanson, William L., 65
Heys, Steven D., xvii, 90
Home-front treatment, 77–78
Horton, Charles H., 68–69
Hospital construction, 101
Hospital ships, 28, 46, 73–74
Hygiene, 100, 128–129

Infections, 13, 20–21, 25, 29–35, 37–38, 41, 67, 91, 98, 122, 129, 131–132, 134
Influenza, 51, 74–77, 100, 126, 128
  numbers, 75
  cases among medical personnel, 76, 97
Injuries, 12, 20–21, 27–28, 30, 33–34, 37, 42–50, 59, 72, 83, 85–86, 98, 101, 106, 110, 112, 116, 119–125, 133, 140

Kosovo, Battle of, 4

Ladd, Anna Coleman, 49, 133
Linker, Beth, xvii, 113, 116
Lister, Joseph, 13, 30

M

Malnutrition, 86
Mayo, Charles, 130, 136–137, 141
McGill, Harold W., 56–59
McIndoe, Archibald, 121
McMeekin, Sean, 4
Medical Corps
 deployment, 54, 56, 60–61, 80–81
 general experience, 54–60
 organization, 22–23, 54–56, 66–67, 128–129, 136, 140
Medical front, 12, 23–24, 137–139, 142–143
Medical knowledge, spread of, 31–32
Medical innovations, in civilian practice, 109–110
Medical personnel
 advancements, 73
 numbers, 81
Medical specialization, xiv, xvii, 13, 25–26, 34, 37, 42, 45, 47, 50, 68, 116–118, 121, 128, 132–133, 137, 140
Medical supplies, delays of, 83
Medical treatment, avoiding, 95–96
Mobilization, armies, 5–6

N

Neurosurgery, 111, 125, 134
Neutrality Acts (U.S.), 142
Nurses, xv–xvii, 13, 55, 60, 62–64, 70–71, 126
Nursing, 70–73

O

Orr, H. Winnett, 65–66, 77, 117, 124, 138

Orthopedics, xiv, 34–38, 65
 after the war, 116–117
 growth of, 117, 124, 135, 138
 hospitals, 68, 102

P

Pain management, 91, 132
Patterson, Robert, 88, 93, 95, 106
Pedicles, use of, 48, 120
Pershing, John J., 9, 79
Physicians, 81, 128, 130–131, 137
 concerns over age, 60–61
 female, 64
 male, 62, 65
 U.S. pre-war, 62
Plastic Surgery, 44–50
 after the war, 117–122
 general treatment, 45
 grafts, 46–47
 difference between U.S. and U.K., 121–122
 modeling casts, 46
Poison Gas, 17–18, 86, 91–93
 treatment of, 73, 93
Progressive era, influence of, 127, 135
Prosthetics, 35, 37, 49, 91, 102, 114–115, 133
Psychological testing, 129
Public health, xvi, 126–128, 130–131, 135–136

R

Race, African-American, 79–80
Rehabilitation, 91, 114–115, 132
 careers after, 112–113
 government's role – 113, 141

174   Index

Reid, Fiona, xvii, 94
Reznick, Jeffrey S., 103, 115
Röntgen, Karl Wilhelm, 40–41
Russia, xv, 5, 9, 11, 120
Russian revolution, 9

S

Sanitary Corps, 128–129
Selcer, Perrin, 51
Serbia, xiii, 3–5
Shell shock, xvii, 104–106, 115–116
   public's view of, 116
Shock, 13, 39–40, 89, 123–124, 140
   prevention of, 124
Skin ailments, 98
Soldiers
   general experience xv–xvi, 7, 18–21
   impressions of medical care, 96–97
   letters, 83–84
   journals/memoirs, xvii, 22, 55, 83–84, 87
   numbers, 1, 11–12
Splints, 34, 36–37, 45, 47, 66–67, 114, 117
Stevenson, Sarah Sand, 71
Stinton, Harry, 87
Stomach issues, 98
Storm, H.H., 101
Stretcher-bearers, xiv–xv, 26, 53, 57, 68–70, 72, 85, 90, 138–139
Surgery, 26–28, 34–35, 41, 67, 89–91, 134

T

Tanks, 15–16
Tetanus, 28–29, 33, 127–128
Todman, Dan, 142

Traction, 35–37, 47, 116
Transport, 26–28, 45, 49, 56, 69, 71, 74, 87–88
Treaty of Versailles, 5, 10, 74, 142
   impact of, 10–11
Trench foot, 20, 86–87
Trenches, building of, 6–7
Triage, 22, 26, 110, 133

U

U-boats, 17, 159
US neutrality, 7–9
US entrance into war, 7–9, 19, 62, 65–67, 113

V

Vaccinations, 127–128
   anti-tetanus, 28, 33, 128
   typhoid, 127
Van Tiggelen, René (*Radiology in a Trench Coat*), 41–42
Venereal disease, 100, 129–130
Veterans, xvii, 12, 102, 112–117, 120, 140
Vocational Rehabilitation Law (U.S.), 102–103
Volunteers, 60–65, 88, 111
   male physicians, 62, 64–65
   numbers, 61
   shortages, 61
   nurses, 62–63, 71

W

"war guilt", 5, 74
Weaponry, 14–15
   new, 17–18

Western Front, xiv–xv, 7, 9–10,
    16, 18–19, 22, 32, 50,
    52, 60, 63–64, 70, 73–74,
    100
Whitehead, Ian, 80
Wilson, Woodrow, 8, 10, 14
World War II, impact on, 11, 14, 79,
    120–123, 135–136, 140
Wounds, closing of, 32–33
Wright, Sir Almroth, 31
World War I as a modern war, 14,
    16–18, 25

X
X-rays, 40–44
    dangers of, 43
    machines, 23, 41
    manual, 42–43
    mobile units, 41–42, 89
    technicians, 42
    uses, 48, 111–112, 137

Z
Zimmerman Telegram, 9

www.ingramcontent.com/pod-product-compliance
Lightning Source LLC
Chambersburg PA
CBHW050139240426
43673CB00043B/1735